CW00496688

CLASSIC WEAPONS SERIES

# THE 88

## THE FLAK/PAK 8.8CM

CLASSIC WEAPONS SERIES

# THE 88

## THE FLAK/PAK 8.8CM

CHRIS ELLIS
AND PETER CHAMBERLAIN

First published in 1998 by
PRC Publishing Ltd,
Kiln House, 210 New Kings Road, London SW6 4NZ

This edition published in 1998 by Parkgate Books Ltd
Kiln House, 210 New Kings Road
London SW6 4NZ



British Library Cataloguing in Publication Data:
A catalogue record for this book is available from the British Library.

ISBN 1 90261 616 2

Printed and bound in China

## PREFACE AND ACKNOWLEDGEMENTS

Very few weapons become legends so quickly as the famous 88 did. It is one of the best known weapons of World War 2 (and was used by some armies who took over ex-German equipment into the 1970s) but its great days were in the 1940-45 period, as recounted and illustrated in this book. Aside from covering the best known models of the 88, however, this book also explains the origins of its evolution and design which can be traced back to World War 1 or even earlier. The many illustrations which also help tell the story come from the Chamberlain Collection, the PRC Collection, the US Army, and Hilary Doyle. The specification is reproduced from the original US Army report on the Flak 36.

A note on nomenclature: the Germans always referred to the weapon as being of 8.8cm calibre, while we have become more used to referring to it as the 88 (Eighty-eight) or 88mm. I have used 88 in narrative or 8.8cm in the text. Rather than putting the Imperial equivalent alongside each time, these are the equivalents for the main calibres identified in the text:

| | |
|---|---|
| 5cm | 1.97in |
| 7.5cm | 2.95in |
| 7.7cm | 3.03in |
| 8cm | 3.15in |
| 8.8cm | 3.465in |
| 10.5cm | 4.13in |

# Contents



PAGE TWO: A classic Western Desert scene with a Flak 18 dug into a defence position against an expected British attack, 18 June 1942. Note the 'kill' bands of enemy tanks destroyed marked on the barrel.

RIGHT: A Flak 36 without armour shield is loaded aboard a transport for despatch to Rommel's Afrika Korps on 25 May 1941.

# Introduction

ABOVE: Classic desert war action: a Flak 36 fires at enemy armour over open sights.

WEAPONS OF WAR that become immediate legends are very few and far between, but the German 8.8cm anti-aircraft gun came into that category in one famous action in 1941, when its effective deployment in a tank battle gave the German Afrika Korps a tactical and morale-boosting advantage far in excess of the number of guns deployed.

The occasion was one of the key incidents early on in the war in North Africa when British and Commonwealth forces had an early clash with the more recently arrived Afrika Korps. Generalleutnant (later promoted Generalfeldmarschall) Erwin Rommel had moved to North Africa via Tripoli in February-April 1941, deploying two Panzer Divisions (21st and 15th) as the Deutsche Afrika Korps to boost the feeble Italian forces which had been swept out of Libya and Cyrenaica by a decisive British and Commonwealth Western Desert Force (later 8th Army) offensive in December 1940 and January 1941. The Afrika Korps was numerically inferior to the Allied forces right from the start, and operations in North Africa were never taken as seriously by Hitler and the German High Command as campaigns in other areas. Rommel made up for his numerical disadvantage by using superior tactics and better handling of his weaponry to wrong-foot the Allies on numerous occasions. The Afrika Korps was constantly short of equipment and supplies, too, which meant that much use was made of captured equipment and supplies.

In early 1941 the Allies were distracted by events in Greece and Rommel took the chance to push back into Cyrenaica, besieging Tobruk and taking up a defensive position at Halfaya Pass on the Cyrenaica-Egyptian border. A British offensive was expected, and in May 1941 Rommel used his characteristic intuition to deploy all 13 of the 8.8cm Flak 18 and 36 guns he had under command, digging them into sangars for use in an anti-tank role. At Halfaya Pass and on Hill 208 they were well concealed, dug in deep, and barely visible from the front. When the expected British attack came early in June 1941 (Operation 'Battleaxe') these few Flak 18 and 36 guns became the perfect surprise weapon. British tanks were feebly armed and poorly armoured, and even the most thickly armoured tank, the Matilda, could be penetrated by the 8.8cm gun. The British tanks were out-ranged and out-gunned, almost literally stopped in their tracks, and 123 of the 238 attacking tanks were destroyed, with further humiliation heaped on the attackers by a skilful counter-attack by the DAK tanks which outflanked the British 7th Armoured Division and nearly cut it off.

An immediate rumour started that the Germans had a 'wonder gun', though the British commanders seemed slow to appreciate what they were up against, for in the next big Allied offensive, Operation 'Crusader', in November 1941, there was a repeat of the carnage of Operation 'Battleaxe' but on a greater scale. By this time the Western Desert Force had been renamed 8th Army and had been greatly expanded. The British fielded 14 tank regiments with over 750 tanks. Against this the German forces, now renamed Panzergruppe Afrika (of which Afrika Korps was a part), had 320 tanks of which 146 were Italian and of dubious value. In Operation 'Crusader', the British attacked with two corps on 18 November 1941, catching Rommel off his guard as he prepared for another attack on the besieged Tobruk. The British tanks attacked in 'cavalry' style and once again they suffered huge losses

against well sited and well dug-in 8.8cm guns. This time the Germans also had 5cm anti-tank guns; they too were superior to the British tank guns. After five days' hard fighting in this battle (Sidi Rezegh) the Germans were down to only about 100 tanks, but the British had lost about 300 out of 450 cruiser tanks deployed, most to anti-tank guns though there were also some mechanical breakdowns. In a counter-attack Rommel personally led his remaining tanks in a sweeping dash for the frontier, outflanking the British in the process.

Even after this defeat the British command was slow to realise the significance of the 88 as a key instrument of balance on the battlefield, though one of the divisional commanders, General Messervy of 4th Indian Division, who experienced the action in the field, certainly attributed the unexpected deployment of the gun in Operation 'Battleaxe' as the main cause of the failure of the attack, and a staff officer after Operation 'Crusader' pointed out that just four 88s stopped a complete British armoured brigade. Rommel was pleased with the results of his tactics. He told a captured British brigadier, 'I don't care how many tanks you British have so long as you keep splitting them up the way you do. I shall continue to destroy them piecemeal.'

The 88 continued to enjoy a formidable reputation for the rest of World War 2. Its success as an anti-tank gun, even though it had been designed as an anti-aircraft gun, soon led to its adaptation as a tank gun and its use in the classic German heavy tanks, the Tiger I and II, only added to its reputation. The Allies struggled to match the 88's hitting power and range and even by the end of the war they had scarcely caught up: it took air power and air supremacy over the battlefield with such aircraft as the rocket-firing British Hawker Typhoon and US Republic P-47 Thunderbolt before German heavy armour could be decisively routed. The power of the 88 was demonstrated time and again. A classic encounter was the action at Hill 213 near Villers-Bocage on 13 June 1944, a week into the Normandy campaign. Here a Tiger tank of 501st Waffen-SS Heavy Tank Battalion, commanded by Obersturmführer Michel Wittmann, decimated an advancing column of the British 7th Armoured Division. Wittman's 8.8cm gun destroyed the leading vehicle and blocked the road; he then picked off all the vehicles behind it, 25 in

ABOVE: A Flak 18 emplaced in a sangar at Halfaya Pass with the recoil cylinder being checked, soon after the June 1941 Operation 'Battleaxe' victory.

ABOVE: Seen at the moment of recoil, this Flak 36 has the later Sondergerät 202 limbers; it has come rapidly into action in the 1943 retreat in Tunis.

RIGHT: A Flak 36 with Sondergerät 202 limbers covers a withdrawal in the Western Desert in May l942; the gun crew is sheltering in hastily dug fox holes.

TOP RIGHT: A classic Western Desert scene with a Flak 18 dug into a defence position against an expected British attack, 18 June 1942. Note the 'kill' bands of enemy tanks destroyed marked on the barrel.

FAR RIGHT: The first tank to be built around the 88 was the Tiger I (Tiger Ausf E) which mounted the 8.8cm KwK 36, a development from the Flak 36. In suitable terrain this slow moving but heavily armed and armoured tank enjoyed some success.

all, while further tanks in the company picked off the rest. The few shots which reached the Tiger tanks from the British vehicles bounced off the thick armour.

While there is no doubt that the 8.8cm gun was a classic design, superbly engineered and of typical German quality, continuously developed and adapted with Teutonic thoroughness, it must also be said that some of its formidable reputation in World War 2 was achieved by the inflexible attitude to its deployment shown by the Allies. The Germans themselves found that the most modern comparable Russian guns they captured were actually superior to the 88 and that the 88 was over-engineered for its task.

It was flexible thinking by the German commanders in general, and Rommel in particular, that made the 88 such a successful weapon. At the time of Operation 'Crusader' a German staff officer summed up the difference between German and British military attitudes: 'A German panzer division is a highly flexible formation of all arms, which always relies on artillery in attack or defence. In contrast the British regard the anti-tank gun as a defensive weapon and fail to make use of their powerful field artillery which should have been used to eliminate the German anti-tank guns.' At this period British staff thinking was still wedded to the tank versus tank battle, a mechanised version of the old cavalry tactics. The notion of drawing tanks onto well concealed anti-tank guns was not part of the thinking, despite the fact that one of the divisional generals taking part in Operation 'Battleaxe', General Creagh, lectured afterwards that, 'When on the defensive the German policy is to draw our tanks on to their guns, then counter-attack with tanks.' This lesson was disregarded when Operation 'Crusader' took place a few months later.

Clearly the fact that British tanks were poorly armoured and under-gunned helped the superiority of the 88 and even smaller German anti-tank guns, for they continually out-ranged and out-performed Allied guns and tanks for most of the war. But the use of the 88 need not have been the surprise it was to the British in the Desert War if previous events had been noted. The Germans first used the 8.8cm anti-aircraft gun in the anti-tank gun role in the Spanish Civil War. More importantly, Rommel himself first used the 88 in the anti-tank role in the Battle of Arras in May 1940, against British tanks, but the significance seems to have been lost on the British at the time. The Battle of Arras was one of the few

bright spots in the inglorious withdrawal of the BEF to the Channel coast during the German Blitzkrieg campaign of May 1940. Rommel commanded the spearhead 7th Panzer Division in this campaign and first came to public fame as a result. A spirited and enterprising counter-attack by the British scratch 'Macforce' stopped 7th Panzer Division's advance and caused a partial German withdrawal. Rommel solved the problem by concentrating all his divisional artillery on the British tanks, and to back up this defence he ordered his 8.8cm Flak battery to fire in the anti-tank role: they knocked out one heavy and seven light tanks from the British force. Two British tanks managed to knock out an 88 gun position in this action.

Ironically the Allies had a weapon available all through World War 2 that was at least the equal of the Flak 88—its British equivalent, the 3.7in anti-aircraft gun, similar in size and performance to the 88, and slightly more powerful. At the time of Operation 'Crusader' the British had more 3.7s than the Germans had 88s, but the 3.7 was classed as an AA gun and British thinking was inflexible on the matter and departmentally-oriented. As it was not officially an anti-tank gun it could not be used in that role, so no attempt was made at command level to copy the German tactics and use the 3.7in gun in an anti-tank role. There were a few occasions when it was unofficially used, but these were rare. On at least one occasion when four 3.7s were available at a critical time in a desert battle the divisional general rejected a suggestion that they be used to fire at approaching tanks and had them moved because they 'got in the way'!

The German 88, then, earned its place and reputation in military history by being the right weapon at the right time, skilfully and resourcefully used and developed. But by the time it became a legend in World War 2 it was already quite an elderly design; its origins went back to the previous world war.

# Early Development

ABOVE: 7.7cm leichte Kw-Flak L/27 on the Mercedes truck showing the sides folded up.

THE GERMANS WERE early to appreciate that defence was necessary against offensive air power. As early as the Franco-Prussian War in 1870 the Prussian Army had what was probably the first purpose-designed anti-aircraft gun. The French were using balloons for observation and to break the siege of Paris. In response to a request Krupp supplied several 25mm guns mounted on pivots on small horse-drawn carts. The gun was hand-operated, traversed by walking round the cart, and elevated by the shoulder. The weapon was successful in that it forced the French balloonists to fly by night to avoid being shot at, but it was not an accurate weapon and lasted only a few months in service. The threat of balloons—airships too after the turn of the century—remained, though military authorities took little serious notice of them.

Possibly with a view to prompting military opinion, and certainly with future sales in mind, two of the leading German armaments firms built 'anti-balloon' guns and demonstrated them at the Frankfurt International Exhibition of 1909. Several different designs were shown by Krupp and Erhardt. Some were on field carriages, but the weapons that made the greatest public impact were on motor lorry chassis. Krupp had a 6.5cm gun with high-angle mount on a 50hp truck, and Erhardt (part of what later became the Rheinmetall company) presented a 5cm gun on two alternative motor carriages, one fully armoured and one open-topped and only semi-armoured. The fully armoured vehicle had the 5cm gun in a traversing turret and full armour protection (3mm nickel steel) for the crew. The 5cm gun was described as a *Ballon Abwehr Kanone* (BAK), meaning balloon defence gun. The Krupp truck-mounted 6.5cm gun was interesting in that the maker had tackled the problem of damping the hammer blow of the recoil that would have had a destructive effect on the chassis. It used a system of differential recoil whereby it was loaded and fired in the recoiled position so that the actual recoil from the moment of firing countered the forward movement from the recoiled starting position. An essential principle was demonstrated by all three of these motorised carriages—fast mobility and all-round traverse of the gun.

No official order, or even trial, of any of these guns came from the German Army, even though a test was carried out that year at the Berlin infantry school where rifles and machine guns were fired intensively at a captive balloon without bringing it down. The official conclusion was that guns larger than rifle calibre would be necessary for balloon destruction from the ground. The 1909 demonstration of the Krupp and Erhardt guns did, however, prompt the British, French and Americans to experiment with air defence weapons of their own. The German Army made only minimal progress in the next few years, although in the 1910-12 period it did acquire the Krupp and Erhardt demonstration vehicles and had two more built, so by the time war broke out in August 1914 the entire German Army air defence consisted of three Krupp guns on Daimler truck chassis and three Erhardt 5cm motorised guns. In addition there were 12 horse-drawn 7.7cm guns. These were used for home defence, and only the six motor guns were sent to protect the field army.

The Krupp motor guns made in 1910-12 had 7.5cm L/35 guns, compared with the 6.5cm gun used on the 1909 prototype. These motor guns were described as *Flugzeugabwehrkanone*, abbreviated to

ABOVE: The Rheinmetall/Erhardt version of the 7.7cm leichte Kw-Flak L/27, showing sides dropped to form a platform for the crew.

*Flak*, meaning aircraft defence gun, probably the first time this name was used.

As soon as the war started it became apparent to all participants that the threat came much less from balloons and airships than it did from aircraft, which had much more offensive power. There was an immediate need for many more anti-aircraft guns and the expedient on the German side came from a large number of 3in calibre Model 1903 field guns captured in the first attack on the Russian front. As large stocks of ammunition were also captured and the gun had high muzzle velocity, they were fitted to pedestal mounts and pressed into service as the 7.62cm *Russische Sockel Fliegerabwehrkanone*. Some captured French 75mm field guns were similarly converted, but these were rebored to take standard German 7.7cm ammunition. Various other ideas were used or tested, including several different ways of using 9cm guns in high angle mounts. Some of these were very crude, using wood chocks to achieve the high angle of fire and turntables to give the traverse. More practical was a further batch of motor guns ordered from Rheinmetall in 1914. These had a purpose-built high-angle pedestal mount for a 7.7cm L/27 gun on an unarmoured Mercedes truck, the sides folding down to form a platform for the gun crew. This was designated 7.7cm leichte Kw-Flak L/27. A similar outfit was ordered from Krupp. Meanwhile the firms of Krupp and Rheinmetall were both asked in July 1915 to come up with purpose-built anti-aircraft guns of larger calibre, namely 8.8cm and 10.5cm.

At the beginning of 1916 the two firms

RIGHT: The 8.8cm K-Zugflak L/45, the first of the 88s. Note twin side outriggers, spoked solid-tyre wheels, and the seat and brake for draft by horse team, though a motor tractor was normal.

BELOW: One of the 8cm Flak L/45 guns built by Krupp along with the 8.8cm and 10.5cm designs.

TOP RIGHT: 8.8cm K-Flak L/45, the Krupp version showing single side outrigger and disk wheels.

BOTTOM RIGHT: Detail of the 8.8cm K-Flak L/45 Krupp gun showing layers' controls.

ABOVE: By contrast to the photograph of the Krupp gun on page 13, this detail shows the layers' controls and right side of the 8.8cm K-Flak L/45, Rheinmetall version.

submitted their designs to the German War Ministry, the Artillery Test Commission and the Inspector of Flakartillerie. Both designs generally met the requested specifications which included high mobility, 360° traverse, elevation of 70°, muzzle velocity (for the 8.8cm gun) of 750m/sec, stabilised four-wheel carriages with pneumatic braking, and towing by motor truck which would carry the ammunition and crew. This proved to be a far-sighted specification that hardly changed for all subsequent Flak 88 developments and made the use of the weapon far more flexible and efficient that the old idea of a truck-mounted weapon. For the 8.8cm gun it was proposed to adapt the 1913 pattern L/45 naval cannon design. This gun was already well proven in naval service and had been successfully tried as an anti-aircraft gun, in its naval mount, by naval units in Flanders in 1915. For trials 8cm calibre versions were also built as well as the 8.8cm and 10.5cm prototypes. The 8.8cm versions were ready by the end of 1916 and were designated 8.8cm K-Zugflak L/45 Krupp and 8.8cm K-Zugflak L/45 Rheinmetall. In this case K-Zugflak stood for towed anti-aircraft gun. While the guns were broadly similar, they differed in mount construction and in carriage style. The Krupp carriage had solid disc wheels and single cruciform stabilisers each side; the Rheinmetall gun had spoked wheels, twin extending stabilisers each side of the carriage, and a horse-dray type driver's seat and handbrake to make it possible for a horse team to draw the equipment where a truck or tractor was not available.

Both the Krupp and Rheinmetall designs were ordered into production and remained in service until the end of the war two years later. Rheinmetall made minor improvements to its design, mainly to the carriage and controls (faster traverse, improved sights) and this resulted in the original design being designated Model 1916 and the improved version Model 1917. General details for both the Krupp and Rheinmetall guns are shown below.

A final Rheinmetall development, Model 1918, had provision for control from linked optical rangefinders, enabling a battery to operate with a degree of centralised control. The earlier guns were operated individually using only the gunsights and local control. By the Armistice of 1918 there were 2,576 anti-aircraft guns of all calibres in the German Army. Most of these were scrapped, however, by the severe terms of the Versailles Treaty. While the German Navy was allowed to keep some AA guns on its warships, the newly established peacetime German army (*Reichsheer*) was allowed no AA guns at all. However some AA guns, with their high angle sights removed and elevation reduced for ground firing only, were allowed to equip the motorised artillery batteries of the Reichsheer's seven artillery regiments. This was, perhaps, a rather unwitting pointer to the future.

| **Data:** | 7.7cm guns |
|---|---|
| **Weight of shot:** | 9.6kg (21.1lb) |
| **Muzzle velocity:** | 785m/sec (2,575ft/sec) |
| **Maximum range:** | 10,800m (35,420ft) |
| **Maximum height:** | 6,850m (22,470ft) |
| **Combat weight:** | 3,010kg (6,622lb) Krupp |
| | 2,785kg (6,127lb) Rheinmetall |
| **Travelling weight:** | 7,300kg (16,060lb) Krupp |
| | 7,200kg (15,840lb) Rheinmetall |
| **Towing speed:** | 12km/hr (7.5mph) |

LEFT: Rheinmetall 8.8cm K-Zugflak L/45 with Erhardt tractor.

Below: Krupp 8.8cm K-Zugflak L/45 with Daimler-Benz tractor. Note provision for a canvas tilt, or possibly a net for camouflage over the gun.

# Between the Wars

ABOVE: Very early production Flak 18 in mid-1930s camouflage finish.

THE CONDITIONS OF the Versailles Treaty of 1919 were understandably very restrictive on the German armaments industry. Among the severe restrictions, Krupp was not allowed to produce guns below 17cm calibre and Rheinmetall was not allowed to produce guns above 17cm, and even then only a few guns a year were permitted.

As these harsh limitations threatened both Krupp's commercial future and its ability to keep abreast of technical development, some cunning ruses were devised to get around the restrictions. In 1921 Krupp made an arrangement with the Bofors company of Sweden whereby Bofors acquired the foreign rights to all Krupp gun designs. In return the Swedes provided design, development and research facilities to Krupp who sent three designers to the Bofors factory. From 1922 the German War Office started providing financial support for this office and a 'bogus' Swedish branch office of Bofors, known as Koch und Kienzle, was set up in Berlin as a cover name for the seconded German design team.

Though popular legend has it that German militaristic expansion did not start until the Nazis came to power under Hitler in 1933, the reality was somewhat different. The Weimar Republic of the 1920s had an army, allowed by the Versailles Treaty, of 100,000, and the C-in-C, General Hans von Seeckt, was an astute and ambitious leader who covertly flouted the treaty limitations where there were loopholes to be exploited, expanded the military force by setting up quasi-military police or border police units, and put forward the idea of a secret agreement with the Soviet Union for building armaments factories and training facilities in Russia for use by the Reichsheer. In the 1925-30 period the Krupp firm (under the Koch und Kienzle name) and Rheinmetall were both asked to forward designs for a 7.5cm heavy anti-aircraft gun for the Reichsheer. This choice of calibre was dictated by a Versailles Treaty clause that allowed the army to have only 7.5cm and 10.5cm guns. Trials on the prototype 7.5cm guns were carried out in 1929-30 but neither proved satisfactory and none were ordered. However, Krupp and Rheinmetall did offer these 7.5cm L/60 guns for overseas sale and they went to such countries as Spain and Brazil. Some that were still in production for export when World War 2 broke out in 1939 were taken over by the German navy (*Kriegsmarine*) and used for coastal defence.

In 1931 the Krupp AA gun designer with the Koch und Kienzle team produced designs for an 8.8cm L/56 gun which was mounted on the same cruciform wheeled carriage that had been produced for the unsatisfactory 7.5cm L/60 gun. The designer took the plans for the new gun to Krupp's Essen factory in 1931 and a prototype was built in 1932. Because of the thorough research carried out under the Bofors 'cover', the new design proved highly successful when tested and it was ordered into production. By the time production was under way in 1933, Hitler was Chancellor and the NSDAP was the party in power, with the old Weimar Republic transformed into the Third Reich and all pretence at abiding by the Versailles Treaty swept away.

The new gun was designated 8.8cm Flak 18 L/56 and proved to be immensely successful. It had the same cruciform platform as the earlier 7.5cm gun, in which the side arms folded up for transport but were folded down to form the transverse section of the cruciform when the gun was

LEFT: Early Flak 18 clearly showing the Übertragungsgerät 30 fire control indicators for elevation and traverse and the elevation (left) and traverse handwheels set below the indicators.

emplaced for action. When emplaced, the cruciform was lowered from the two two-wheel axle units which provided the mobility. When emplaced, the gun had an all-round traverse. It was a semi-automatic weapon which ejected the spent cartridge on recoil and re-cocked itself for the next round which could either be loaded by hand or with a power-assisted rammer. This enabled a well-trained crew to keep up a very good rate of fire. The range of 8,000-10,000m was adequate against aircraft then in service.

The gun had a conventional one-piece barrel which fired a high explosive (HE) round with a conventional copper driving band. At the rate of fire which could be achieved the barrel was good for firing about 900 rounds before being replaced.

The procurement authority—the Army Weapons Office, (*Heereswaffenamt*), considered this posed logistic problems in the supply of replacement barrels in the field and the disposal of old ones. Rheinmetall solved this problem by designing a three-piece barrel, designated RA9 (*Rohr Aufbau*—barrel structure—9). This featured a jacket, a sleeve and an inner three-part tube, of which the start of the rifling and the forcing cone was in the centre section. Most barrel wear takes place in this part of the tube, so for field replacement in most cases only this section needed to be replaced. There was a locking collar holding the breech to the jacket and the three-section inner tube was held by clamping rings front and rear. It was much easier to strip the gun while emplaced and replace this centre section than it was to bring up complete barrels and their handling equipment. For the most part only centre sections had to be carried as spares.

There was a price to pay for this clever idea, but at the time it was considered a necessary penalty for combat efficiency. The three-section barrel had to be made of the finest quality steel and very fine tolerances were necessary to get a perfect fit at the joins, so production time and cost was in excess of the conventional one-piece barrel. Also the three-part barrel was heavier, so the recoil and equilibriator mechanisms had to be beefed up to compensate. The new three-part barrel was finalised in

ABOVE: Early Flak 36 fitted with armour shield and the original Sondergerät 201 limbers. Note the traverse leaf spring suspension and the hand winch and chain which raises and lowers the cruciform carriage in emplacement.

BELOW: The Sd Kfz 8 12-ton tractor was a type used to haul 8.8cm guns, particularly the later types such as the Flak 41 or Pak 43.

1936 and put into production in 1937. Several further improvements were made to the equipment, including a cruciform and new limbers (wheeled bogies) of stronger construction designed by the firm of Linders. The changes were intended to improve mobility and stability in action and simplify mass production of the equipment. The new limber was designated *Sonderhanger* 201 (special suspension 201), and it was designed in response to early experience in the Spanish Civil War in 1936 where the German volunteer Condor Legion fought on the Nationalist side. The new Sonderhanger 201 speeded up time needed to bring the equipment into action. A further development in 1939 improved further on the carriage with the Sonderhanger 202, which enabled the gun to be fired while still on its wheels. In view of the key part the weapon was to take in anti-tank warfare, this was another important step forward.

Though the Flak 18 was sent with the Condor Legion as an anti-aircraft gun, it was in Spain that it was increasingly used in an indirect fire role and even in the direct fire anti-tank role. It is said that Hitler himself suggested trying it against tanks in 1938. Perhaps because the Spanish Civil War did not directly involve armed forces of the big powers, the potential of the Flak 18 as an anti-tank gun was not appreciated by some of them, particularly Britain and France. Armour-piercing ammunition for the 8.8cm gun was, accordingly, developed in 1938 and Spanish Civil War experience also led to the introduction of a gun shield.

The new version of the gun, with three-part barrel and the Sonderhanger 201 limbers, was designated 8.8cm Flak 36 L/56 and entered production and service in 1937. Both the Flak 18 and Flak 36 carried a fire control transmission system designated Übertragung 30. This linked the guns to the predictor and rangefinder. With the Übertragung 30 the receiver dials had three concentric circles of bulbs and three mechanical pointers pivoted centrally. The appropriate bulbs lit up to indicate the data being transmitted and the rate operator on the gun moved the pointers to cover the lights, thus giving a continuous transmission of elevation, traverse and deflection to engage aircraft. A much improved Übertragung 37 transmission system was developed in 1937 and was in production in 1939. The Übertragung 37 was a Selsyn system, whereby the layer on the mounting followed and matched the pointers to moving pointers indicating changing data sent through from the predictor operator. This was more efficient and faster in transmission than the Übertragung 30 and the Flak 36 was modified to take it. In this form the equipment was designated 8.8cm Flak 37 L/56. The improved Sonderhanger 202 limber was also introduced in 1939.

The Flak 18, 36 and 37 all had the same ballistic performance and differed

ABOVE: The standard towing vehicle for the Flak 18, 36 and 37 was the Sd Kfz 7, here seen in prewar style camouflage in 1938. This was an 8-tonner.

LEFT: Emplacing a Flak 18 in the early days of the desert war. The carriage has been lowered and the limbers are being wheeled clear while the outriggers are lowered to the ground at the sides.

from each other only in the physical and constructional changes outlined above. In practice the differences between them often became less than clear cut. With the inevitable sequence of repairs, overhauls and updates, it was possible to see the various features retro-fitted between the different models. Thus the one-piece barrel of the Flak 18 could be found on the Flak 36 and 37, and the three-part barrel could be found on the Flak 18. The gun shield could be fitted to any of them, or omitted, and the Sonderhanger 201 and 202 could be found on the Flak 18. Even the Übertragung 37 could be found fitted to the earlier Flak 18 and 36, strictly speaking converting them to Flak 37s. In addition to the anti-aircraft sights and controls the weapons were fitted with telescopic sights for anti-tank work and later in the war some had the transmission systems removed or isolated so they were then only suitable for anti-tank fire. The Flak 18, 36 and 37 were in service throughout World War 2.

| **Data:** | Flak 18, 36 and 37 |
|---|---|
| **Calibre:** | 8.8cm |
| **Length of barrel:** | 4.9m (16ft 2in) |
| **RH rifling length:** | 4.1m (13ft 6.3in) |
| **No of grooves:** | 32 |
| **Traverse:** | 2 x 360° |
| **Elevation:** | From -3° to +85° |
| **Recoil at 0°:** | 1.05m (3ft 5.34in) |
| **Recoil at 25°:** | 0.85m (2ft 9.46in) |
| **Recoil at 85°:** | 0.7m (2ft 3.75in) |
| **Recoil at max elevation:** | 1.08m (3ft 6.5in) |
| **Transmission:** | ÜTG 30 (Flak 18 and 36), ÜTG 37 (Flak 37) |
| **Firing system:** | Percussion |
| **Rate of fire:** | 15-20 rpm |
| **Ceiling (maximum):** | 10,600m (34,770ft) |
| **Ceiling (effective):** | 8,000m (26,250ft) |
| **Carriage:** | Cruciform with Sonderhanger 201 or 202 limbers (also produced with static defence and railway mounts) |
| **Overall length:** | 7.62m (25ft) |
| **Overall width:** | 2.3m (7ft 7in) |
| **Overall height:** | 2.42m (7ft 11in) |
| **Overall weight (mobile):** | 6,861kg (15,129lb) |
| **Fuse setter:** | Zunderstellmaschine 18 (Flak 18 and 36), Zunderstellmaschine 19 or 37 (Flak 37) |

**Ammunition data**

| *Projectile* | *Muzzle velocity* | *Weight of round* |
|---|---|---|
| HE time | 2,690ft/sec (820m/sec) | 20lb 1oz |
| HE percussion | 2,690ft/sec (820m/sec) | 20lb 5.5oz |
| APCBC | 2,600ft/sec (795m/sec) | 21lb 0.5oz |

**Gun detachment**

| | |
|---|---|
| Commander | 4 & 5. Ammunition supply |
| Tractor driver | 6. Fuse setter operator |
| 1. Layer for elevation | 7. Fuse setter (round handler) |
| 2. Layer for line (traversing) | 8. Ammunition supply |
| 3. Loader | 9. Ammunition supply |

LEFT: Flak 18 of a divisional Flakartillerieregiment on the move in 1939. Note Luftwaffe number plates and spare wheels carried on the Sd Kfz 7 tractor.

BELOW LEFT: An early Flak 18 with armour shield but without side extensions is hauled by a semi-armoured half-track during the campaign in France, 31 May 1940. This armoured version of the Sd Kfz 7 (Gepanzerte Zugkraftwagen 8t) was issued to panzer divisions in 1939-40.

BOTTOM LEFT: A Flak 37 showing the Übertragungsgerät 37 fire control 'follow the pointers' indicators replacing the earlier type.

# Wartime Developments

ABOVE: Flak 41 emplaced with gun at maximum elevation.

AS THE POLITICAL developments in Europe made war more likely, the main powers rearmed. The new aircraft that were being developed had a vastly better performance than military aircraft of the early 1930s—they could also fly higher. Therefore in 1938-39 the German War Ministry decided that better performance anti-aircraft guns were required. An obvious answer was to develop new 10.5cm and bigger pieces, but it was considered that the 8.8cm gun could also be improved, most importantly with a much higher muzzle velocity. This was considered achievable due to the new development of Gudol propellant and sintered iron driving bands. Battle experience in Spain led to the intention of making the new design suitable for both the anti-aircraft and the anti-tank role right from the start. For the latter role it was made much lower, with a turntable type chassis, and to make it lighter there was to be no power mechanism. The new design was originally given the designation Gerät 37 and Rheinmetall was given the contract in autumn 1939 to develop it. This designation was very similar to Flak 37, and to prevent confusion the official designation was changed to 8.8cm Flak 41 in summer 1941.

## Flak 41

The new gun had a long L/74 barrel that was made in five parts rather than the three parts used for the Flak 36 and 37. This added to production costs and time. Later a three-part barrel was developed, and in the last year of the war and too late to have much effect, a two-part barrel was developed for all the different 8.8cm guns.

The prototype of the Rheinmetall Flak 41 was ready for trials in summer 1941, but it suffered from teething troubles, the main one of which was never really satisfactorily overcome. This was jamming of the breech and the difficulty of extracting cartridges due to uneven expansion of the multi-section barrel and the use of steel instead of brass cartridge cases as an inevitable war economy. There was no problem when brass cartridge cases were used but the supply of these was haphazard. In spring 1942, when the trials had been completed, Munitions Minister Albert Speer wanted to cancel the project due to its complexity, cost and less than satisfactory trials. However, he was over-ruled by Hitler who ordered the gun into production. A first batch of 44 guns was built for combat trials and these were sent to North Africa. They proved unreliable and spent much time under repair. When the German forces evacuated North Africa in 1943, they left these behind, although took their earlier 8.8cm guns with them. Faults and all, however, the Flak 41 remained in production until the war's end. Apart from the extraction and jamming problems, the Flak 41 was actually superior to the earlier Flak 18, 36 and 37 models. The turntable improved the traversing speed and the low silhouette made concealment much easier.

The muzzle velocity was increased by about 180m/sec (600ft/sec). The loading mechanism was very ingenious. The drive for this was obtained by fitting an auxiliary hydro-pneumatic 'recuperator' gear. The piston was withdrawn to the rear during recoil and held withdrawn until released by a catch when the round was placed in the breech opening. On release the rod was taken forward by the increased air pressure, reasserting itself in the 'recuperator', and teeth on the rod rotated some rubber rollers which gripped the round and pulled it into the chamber.

| Data: | 8.8cm Flak 41 |
|---|---|
| Calibre: | 88mm |
| Length of gun: | 6.55m (21ft 5.75in) |
| RH rifling length: | 5.41m (17ft 9in) |
| No of grooves: | 32 |
| Traverse: | 360° |
| Elevation: | From -3° to +90° |
| Recoil at 0°: | 1.2m (3ft 11in) |
| Recoil at 90°: | 0.9m (2ft 9.25in) |
| Firing system: | Electric |
| Transmission system: | ÜTG 37 |
| Fuse setter: | Zunderstellmaschine 41 |
| Rate of fire: | 15 rpm |
| Ceiling maximum: | 15,000m (49,200ft) |
| Carriage: | Mobile turntable and Sonderhanger 202 limbers |
| Overall length: | 9.66m (31ft 8in) |

**Ammunition data**

| *Type of projectile* | *Muzzle velocity* | *Weight of projectile* |
|---|---|---|
| HE | 1,000m/sec (3,280ft/sec) | 20lb 11oz |
| APCBC | 980m/sec (3,214ft/sec) | 22lb 8oz |

**Gun detachment**

As for Flak 18, 36 and 37 but with one additional man as assistant gun layer. Total: 12

TRAVELING POSITION

Recuperator rod disconnected so that barrel may be retracted for traveling

ABOVE AND BELOW: Flak 41 secured for towing and emplaced. *US Army diagram*

### Gerät 42

In spring 1941 Krupp was also asked to produce a new 8.8cm gun to the Gerät 37 requirement but, after experience with the Rheinmetall design, the muzzle velocity was to increase to 1,020m/sec (3,608ft/sec) and the shell weight to 10kg (22lb). In view of these changes the project was designated Gerät 42. The project was never realised in completed form, however, for it got submerged in subsequent requirements for the development of a 8.8cm tank gun and the Pak 43 anti-tank gun. It was envisaged, however, that the designation would have been 8.8cm Flak 42 L/80 and that the ammunition would have been standardised with that for the tank gun and the Pak 43. The prototype was scheduled for completion in spring 1943. The muzzle velocity was finalised as 1,100m/sec (3,610ft/sec) with a round 1.22m (4ft) long. In February 1943, however, the Gerät 42/Flak 42 project was cancelled in favour of the more important tank gun and Pak 43 work.

BELOW: The 8.8cm Kanone auf Sfl project was originally to feature the Gerät 42, but it only existed as a wooden mock-up as the project was cancelled. Note the muzzle brake.

ABOVE: The Flak 41 emplaced on its cruciform.

RIGHT: The 8.8cm Pak 43 on its wheeled carriage from which it could be fired in limited traverse.

ABOVE AND LEFT: The 8.8cm Pak 43 emplaced on its cruciform which allowed a 360° traverse and gave it a very low height.

## The Flak 37/41

When the Flak 41 went into production in spring 1942 it was realised by the German Air Ministry (responsible for home air defence) that it would be at least two years before sufficient were available to replace the older Flak 18, 36 and 37 in the anti-aircraft role. To supplement the 10.5cm AA guns it was decided to try to marry the new Flak 41 gun to the carriages of the Flak l8, 36 or 37. But because of the slow production rate, there were insufficient spare Flak 41 barrels to make this a practical proposition and this was aggravated by the extraction and jamming problems already noted. In addition the old carriages would not stand up to the stresses imposed by the firing of the Flak 41 gun when the rig was tested. The solution was to fit a muzzle brake which was done by fitting an extra length on the barrel with holes bored in it and two flangeplates screwed on the outside. More powerful springs were fitted in the equilibriators. The fuse setter and rammer used on the Flak 41 were also fitted. There were further technical problems, however, and the extraction problem was never solved. Though a small number of these new guns were produced under the designation 8.8cm Flak 37/41 l/80, it is thought that it was never approved for full production. One measure included modifying the Flak 18 or 36 barrel but this also proved to be complicated.

**Data:** 8.8cm Flak 37/41. Details generally as for Flak 31 (gun) and Flak 37 (carriage) except—

| | |
|---|---|
| **Length of barrel and breech:** | 7.03m (23ft 0.6in) |
| **Length of rifling:** | 5.85m (19ft 2.5in) |
| **Elevation:** | From -5° to +85° |
| **Rate of fire:** | 16-20 rpm |
| **Maximum ceiling:** | 15,000m (49,200ft) |
| **Weight of equipment:** | 7,111kg (7 tons) |

**Ammunition**

| *Type of projectile* | *Muzzle velocity* | *Weight of projectile* |
|---|---|---|
| HE | 1,000m/sec (3,280 ft/sec) | 20lb 11oz |
| APCBC | 980m/sec (3,214ft/sec) | 22lb 8oz |

## Pak 43

While Krupp's Gerät 42 (Flak 42) project was abandoned before completion, the work was not wasted since it was carried out against the background of an attempt to produce a new 8.8cm family of guns and ammunition for universal adaptation for fitting to tanks and self-propelled artillery as well as on mobile mounts. The most important design to emerge after the cancellation of the Gerät 42 was a version of the 8.8cm designed solely as an anti-tank gun. It was one of the best and most effective anti-tank designs of World War 2.

As with the early anti-aircraft gun designs, it was on a cruciform carriage with two bogie wheel sets, and it could, in an emergency, be fired from the carriage although traverse in this case was limited to 30° each side. The gun was normally emplaced on its cruciform which gave it a 360° traverse. With a good sloped armoured shield, it was very low when emplaced, standing only about 135cm (4ft 6in) high. An increased range of armour-piercing ammunition was developed for

RIGHT: A well emplaced 8.8cm Pak 43 in action in France in summer 1944, being used by the US troops who captured it against its former owners. Note the bogies hauled clear of the cruciform in the foreground.

this gun and it could knock out any Allied tank including the heavily armoured Soviet KV and Stalin tanks. There was a direct fire sighting telescope and an auxiliary sight for indirect fire. A clever feature was an electrical cut-out to prevent the gun firing and recoiling at high angles if the breech was above one of the cruciform members at the time. Another feature to accommodate the low height of the piece was a semi-automatic vertical falling breech block. The barrel was relatively lightly made, which meant that, when firing AP rounds, the barrel life was quite short, only about 500 rounds. This was the only major drawback of an otherwise superb design.

ABOVE: The Pak 43/41.

| | |
|---|---|
| **Data:** | 8.8cm Pak 43 L/71 |
| **Length of barrel:** | 6.61m (21ft 8.23in) |
| **RH rifling length:** | 5.13m (16ft 9.75in) |
| **No of grooves:** | 32 |
| **Chamber length:** | 0.86m (2ft 9.88in) |
| **Traverse:** | 360° (emplaced), 30° each side (on carriage) |
| **Recoil maximum:** | 1.2m (3ft 11.25in) |
| **Overall length:** | 9.2m (30ft 2in) |
| **Overall height:** | 2.05m (6ft 9in) (on carriage) |
| **Overall width:** | 2.2m (7ft 2.5in) |
| **Overall weight:** | 5,000kg (11,025lb) |

**Ammunition data**

| *Type of projectile* | *Muzzle velocity* |
|---|---|
| HE KwK* 43 | 700m/sec (2,298ft/sec) |
| HE 43 KwK 43 | 750m/sec (2,460ft/sec) |
| AP 39-1 KwK 43 | 1,000m/sec (3,282ft/sec) |
| AP 39/43 KwK 43 | 1,000m/sec (3,282ft/sec) |
| AP 40/43 KwK 43 | 1,130m/sec (3,708ft/sec) |
| HE 39 HL Kwk 43 | 600m/sec (1,968ft/sec) |
| HE 39/43 HL KwK 43 | 600m/sec (1,968ft/sec) |

* KwK=*Kampfwagen Kanone*=armoured vehicle gun)

**Typical armour penetration**

*With 8.8cm AP 39/43*

190mm (vertical) 167mm (30°) at 1,000 yards

145mm (vertical 127mm (30°) at 2,500 yards

*With 8.8cm AP 40/43*

241mm (vertical) 192mm (30°) at 1,000 yards

159mm (vertical) 114mm (30°) at 2,500 yards

RIGHT: All round views of the Pak 43/41, with and without trails opened for emplacement.

ABOVE: Breech detail of Pak 43/41.

**Pak 43/41**

The success of the Pak 43 led to demand for more, but production could not keep up with demand, partly due to the disruption caused by Allied bombing of armaments factories. There were more gun barrels available than carriages for the Pak 43, so a quick expedient was to mount the gun on an older conventional type of field gun carriage. Various old carriages were tried and found wanting. Eventually the gun carriage used was that intended for the 10.5cm le.FH 18 (*leichte Feldhaubitze*/light field howitzer) which was obsolete. To reduce its height the wheels were taken from the 15cm s.FH 18 (heavy howitzer) and a new shield was provided. A dial sight was added to allow the piece to be used in the field gun role as well as the anti-tank role. The gun was the same as the Pak 43 except that the breech was simplified. The full designation was 8.8cm Pak 43/41 L/71. As it stood higher than the Pak 43 and was nose heavy it was not such a good piece to handle, but it was a most effective stop gap in the desperate last year of the war.

**Data:** 8.8cm Pak 43/41 L/71. Details as Pak 43 except—

| | |
|---|---|
| **Length of gun:** | 6.62m (21ft 8.23in) |
| **Overall length:** | 9.14m (30ft 1in) |
| **Overall width:** | 2.53m (8ft 3.5in) |
| **Overall height:** | 1.98m (5ft 8in) |
| **Overall weight:** | 4,380kg (9,656lb) |
| **Elevation:** | From -5° to +38° |
| **Traverse:** | 56° |

**Captured Russian Guns**

The German invasion of Russia in June 1941 had tremendous initial success and resulted in the capture of vast amounts of equipment. Three standard Russian anti-aircraft gun types were included in some numbers, the 7.62cm Model 1931, the 7.62cm Model 1938 and the 8.5cm Model 1939. Much ammunition suited to these weapons was also captured and the guns were first used by the Wehrmacht unaltered. Many were sent back to Germany to supplement AA defences and there were sufficient weapons available for printed handbooks to be produced by the ordnance department. The Model 1931 gun was copied from the Vickers AA gun of that era and was the least satisfactory, mainly due to a poor carriage and iron wheels. The Model 1938 gun was developed as a result of Spanish Civil War experience and was found to have an excellent performance and be very mobile. The Model 1939 was an improved gun achieved simply by fitting a larger calibre 8.5cm barrel. This gun was so impressive that it was thoroughly examined and tested by the ordnance department, whose report stated that, while the Soviets had not used any original ideas, they had copied the best features of others and that with the 8.5cm gun they had produced a weapon virtually as good as the 8.8cm Flak 36 and 37 but of lighter and simpler construction, producing more efficiency at less cost. Adaptation of the 7.62cm Model 1938 to the 8.5cm Model 1939 was particularly effective, achieving a superior weapon at minimal development cost. A muzzle brake on the 8.5cm gun was used to make the same mount hold a more powerful gun. The report suggested that the German arms makers could learn lessons from the Russian designs. While it acknowledged that the Russians worked to lower safety standards, there seemed to be some scope for studying the lighter construction in order to achieve production economies and chamber modifications to German designs. The report concluded that the 8.5cm Model 1939 was at least the equal of the Flak 36 and 37 and was in some respects superior, hence there was no reluctance to take these guns into German service.

When the captured Russian ammunition ran out in 1942 it was decided to modify the captured guns to take the 8.8cm ammunition used in the Flak 18, 36 and 37. This was achieved either by reboring the Russian barrels to 8.8cm calibre or using the 8.8cm Flak Seelenrohr barrel. So modified it was possible for the Russian guns to use the German 8.8cm ammunition and the related German firing tables. The modified guns were allocated both to Luftwaffe flak units and to army units in the field and they were cleared for both anti-aircraft and anti-tank use. A shield was also fitted to some of the Model 1938/39 guns used by the army in the field. The first 71 8.5cm guns converted to 8.8cm (Model 1939) were listed as being in service in October 1942 and priority was given to these as the Model 1931 and Model 1938 conversions are noted as only just beginning at the end of 1942. Though some reboring work was carried out by an ordnance works in North Italy, records show that very many were done by Rheinmetall in the 1942-44 period. Very few of the Model 1931 guns were rebored and used but the Model 1938 and 1939 guns were used to the end of the war. In German service these guns were designated 7.62cm Flak M.31(r), 7.62/8.8cm Flak M.38(r) and 8.5/8.8cm Flak M.39(r) respectively.

### Flak 39/41

A further project was to mount the Flak 41 gun on the carriage of the 10.5cm Flak 39. The resulting hybrid was designated 10.5cm Flak 39/41 L/74. However, only test prototypes were made of this and it never entered production. No doubt the shortage of Flak 41 barrels and the same expansion and extraction problems associated with this gun contributed to the abandonment of the project.

| | |
|---|---|
| **Data:** | Flak M.38(r) and Flak M.39(r) as rebored to 8.8cm calibre |
| **Weight:** | 4,210kg (9,262lb) M.38<br>4,200kg (9,240lb) M.39 |
| **Length overall:** | 5.15m (16ft 10.7in) |
| **Width on carriage:** | 2.25m (7ft 4.5in) |
| **Width emplaced on cruciform:** | 4.8m (15ft 9in) |
| **Height:** | 2.22m (7ft 3.4in) |
| **Weight emplaced:** | 3,047kg (6,703lb) M.38<br>3,057kg (6,725lb) M.39 |
| **Length of barrel:** | 4.2m (13ft 9in) M.38<br>4.7m (15ft 2in) M.39 |
| **Traverse:** | 360° |
| **Elevation:** | From -3° to +82° (M.38)<br>-2° to +82° (M.39) |
| **Gun detachment:** | 11 |

BELOW: Pak 43/41 in a camouflaged emplacement in Italy, ready to engage target.

TOP: The old carriage of the Pak 43/41 made it heavy. Here the crew push the gun into place in a training exercise.

ABOVE: Flak 7.62cm M38(r) as first used by the Germans. Later they were rebored to 8.8cm.

RIGHT: Flak 8.5cm Flak M39(r)—best of the captured guns, most were rebored to 8.8cm when captured ammunition was used up.

FAR RIGHT: Pak 7.62/8.8cm Flak M31(r), one of the captured and rebored Russian types.

# Tank and Self-propelled mountings

ABOVE: Frontal view of the biggest German AFV to mount the 8.8cm gun—the Tiger II King Tiger which had the KwK 43 L/71. This is the Porsche-designed turret.

IT WAS HITLER, after watching a demonstration of the new Flak 36 in 1938, who first suggested that the gun could be used in an anti-tank role, and also that it should be mounted in a tank. At the time there was no tank in German service big enough to take a 8.8cm gun, but the Henschel company at that time was already working on new designs for a prototype intended to replace the Panzerkampfwagen IV, then the heaviest German tank in production and equipped only with a 7.5cm howitzer. Originally the Henschel design was to be only 30-33 tons with the same 7.5cm gun as the PzKpfw IV, but this requirement was soon changed, several times, to bring the weight up to 36 tons and the armament to a 7.5cm L/48 high velocity gun. The appearance of the Soviet T-34 tank in 1941, with its very good 7.6cm gun, made this project obsolete, and the requirement was changed to an even larger tank, nominally 45 tons with an 8.8cm gun. Dr Ferdinand Porsche, a favourite of Hitler's, had been asked to produce another design to the same specification as a safeguard against failure of the Henschel design. By this time, four years of changing requirements had slipped by and the situation required urgent results. It was stipulated that the Henschel design, then designated VK4501, be ready for demonstration on Hitler's birthday, 20 April 1942. The Porsche design was to be shown at the same time, but both prototypes carried the same Henschel-designed turret and the same 8.8cm armament.

The Henschel design was judged the best and the new tank was ordered into production starting in August 1942. This became the famous Tiger tank, designated Panzerkampfwagen VI Tiger Ausf H, Sd Kfz 181. The designation was later changed to PzKpfw Tiger Ausf E, Sd Kfz 181. This was the first tank in service with the 8.8cm gun. It was a massive vehicle, ending up with a weight of 56 tons rather than the 45 tons originally specified. It had 100mm of frontal armour, wide tracks, torsion bar suspension and interleaved triple overlapping disc wheels, originally rubber-tyred, but steel-tyred and simplified on later production vehicles. The tank was built to high quality and was expensive and complex to make. It was first in action in Russia, with no great distinction, in 1942, and was also sent to Tunisia early in 1943 in the final months of the North African campaign. It remained in production until August 1944 and 1,350 were built.

The 8.8cm KwK 36 gun was derived from the Flak 36. The principal modifications were the addition of a muzzle brake and electric firing by a trigger-operated primer on the elevating handwheel. A 7.92mm MG34 was coaxially mounted on the left side of the mantlet. The KwK 36 had a breech of the semi-automatic falling wedge type, scaled up from that used on smaller German tank guns. The great weight of the barrel was counter-balanced by a large coil spring housed in a cylinder on the left-hand front of the turret. Elevation and traverse were controlled by handwheels to the right and left of the gunner respectively and the commander had an additional traverse handwheel for emergency use.

Because of the turret and gun's great weight, traverse was necessarily low geared in both hand and power. It took 720 turns of the handwheel to traverse through 360°. This was a tactical disadvantage for the Tiger and often gave the enemy a chance to get in the first shot—or take avoiding action—while the big turret was slowly traversed to bear. However, the thick armour and high power of the 8.8cm KwK 36 gun gave the Tiger superiority in a direct shooting match or in a good ambush position.

The KwK 36 differed from the Flak 36 in having a one-piece barrel and it had entirely different sighting arrangements with a TZF 9b binocular telescope and an auxiliary clinometer for use in indirect laying. There was also a stereoscopic telescope in the turret cupola and a coincidence rangefinder in the turret roof, plus a sighting vane in the cupola episcope. Some 92 rounds of HE and AP ammunition was carried in the vehicle. Such was progress in wartime, however, that the Tiger was the only tank to carry the KwK 36 gun.

| **Data:** | Kwk 36 L/56 gun |
|---|---|
| **Length of gun:** | 4.86m (16ft 2.3in) |
| **Length of rifling:** | 4.03m (13ft 5.1in) |
| **Length of chamber:** | 0.59m (1ft 11.6in) |
| **Length of muzzle brake:** | 0.38m (1ft 3.1in) |
| **Overall length:** | 5.235m (17ft 5.4in) |
| **Weight:** | 1,331kg (2,932lb) |

**Ammunition data**

| *Type of round* | *Muzzle velocity* |
|---|---|
| APC 38 | 810m/sec (2,657ft/sec) |
| HVAP 4 0 | 915m/sec (3,000ft/sec) |
| HEAT 39 HI | 600m/sec (1,968ft/sec) |
| HE | 820m/sec (2,690ft/sec) |
| **Total rounds carried:** | 92 (in varied combinations) |

### 8.8cm Flak 18(Sf) auf Zugkraftwagen 12t, Sd Kfz 8

The very first attempt to produce a mobile self-propelled version of the 8.8cm gun was a simple project to fit a Flak 18 on the chassis of the Daimler-Benz DB10 ZgKw 12-ton half-track tractor. These were intended to act as both anti-tank vehicles and also for destroying fortified positions. The gun was bolted to the floor in the rear cargo space of the vehicle with the side outrigger arms removed. The barrel faced forward over the armoured hood of the vehicle, and an armoured cab was provided for the driver. The gun had an armour shield with extended sides. Ten of these conversions were produced in 1939 and another 15 were completed in 1940. The overall weight was about 17 tons and these vehicles were issued to 8th Heavy Tank Destroyer Unit (*8.Schwere Panzerjagdabteil*) and were used in the French campaign of May/June 1940 .

### 8.8cm Flak 37 Selbsfahrlafette auf 18t Zugkraftwagen

The next self-propelled 8.8cm gun on a semi-track chassis came as a result of requests from North Africa for heavier weapons. The Heeres Waffenamt produced in response a project that they had worked on some time previously. This was an 8.8cm Flak 37 fitted on the chassis of the big Daimler-Benz 18-ton heavy half-track. Armour protection was added over the engine and crew position. The first vehicle was completed for trials on 31 October 1942. The intention was for the equipment to have a dual anti-aircraft and anti-tank role and it was fitted and provided with ammunition accordingly. Pzgr. and Pzgr.

BELOW: A late production Tiger Ausf E on the Russian Front, showing the big KwK 36 8.8cm gun and its heavy frontal armour.

RIGHT: The 8.8cm Flak(Sf) auf Zugkraftwagen 12t was the first SP version of the 88, here shown in France June 1940.

BELOW: The 8.8cm Flak 37 Sf auf 18t Zugkraftwagen had a dual AA and A/T role but only 12 were built in 1943.

BELOW RIGHT: Versuchsflakwagen 8.8cm Flak (Pz Sfl IVc) as originally completed with the Flak 41 mount, shown with armoured sides raised.

39 APCBC/HE rounds were provided for anti-tank use, while the Flak 37 could also engage aircraft flying at 3,000-5,000m (9,840-16,400ft). Though 112 vehicles were ordered after trials there were other priorities by this time and only 12 were built in the July-September 1943 period. They were issued to 2nd Schwere Batterie/Heeres Flakartillerie-Abteilung (Sf) 304 which was sent to Italy in September 1943 and became part of 26th Panzer Division.

### VFW 8.8cm Flak auf Sonderfahrgestell (Pz Sfl IVc)

Another early attempt to mount the 8.8cm on a mobile chassis arose from the French campaign of 1940. Three Pz Sfl IVc mounting 8.8cm L/56 guns were requested for the task of knocking out the heavily fortified bunkers of the Maginot Line. The timescale was against this since the Maginot Line was bypassed and the campaign was over in weeks. After the fall of France it was suggested that they could be used as tank destroyers instead, but Krupp, who was developing the design, pointed out that they would be too lightly armoured for this role. There were considerable problems with the development of the new chassis, particularly in finding a reliable transmission and steering system. The chassis running gear had interleaved wheels and was effectively a combination of PzKpfw IV and half-track chassis components.

In June 1942 Krupp suggested that the chassis should be used as the basis of an anti-aircraft tank (*Versuchsflakwagen*—VFW—experimental AA tank) mounting an 8.8cm L/71 gun. Two were ordered for trials to be completed in April and May 1943 but completion was delayed, partly by more transmission problems but also due to the bombing of Krupp's Essen factory which necessitated transfer of assembly to Magdeburg where the prototype was finally completed in November 1943. In the meantime the order for the second vehicle was cancelled and the parts for it were scrapped. In January 1944 armaments minister Speer ordered cancellation of the project as the concept of having a large calibre Flakwagen escorting tanks had been rendered obsolete in favour of small-calibre vehicle-mounted guns which could engage low flying strafing aircraft. It was also realised that effective fire control would be impossible with moving vehicles as cables were still needed to link the predictor to the gun. A Flak 41 was fitted to the chassis and used for firing trials in Denmark in March 1944. After this a Flak 37 replaced the Flak 41 and the vehicle was finally sent to Heeres Flakartillerie-Abteilung (Sf) 304, part of 26th Panzer-Division, in Italy for field trials. This vehicle was neatly designed, having armoured sides and ends which could be lowered to form a platform for all-round traverse.

This vehicle was also known at one stage as the Grille 10 (see page 44).

the barrel was held down by a stay on the vehicle nose. The vehicle was first used in Russia, then Italy and finally in the West. The name Hornisse was later changed to *Nashorn* (Rhinoceros).

**Sturmgeschütz mit 8.8cm Pak 43/2 Sd Kfz 184 Ferdinand/Elefant**

When the VK4501 project for a heavy tank was projected both Henschel and Dr Ferdinand Porsche were asked to produce designs. Porsche was a favourite of Hitler and 90 of the Porsche VK4501 design were ordered 'off the drawing board' as a safeguard against failure of the Henschel design. Only two of the Porsche tanks were actually completed as tanks (with Henschel turrets) and one of these took part in the April 1942 trials against the Henschel design. A KwK 36 8.8cm tank gun was intended for this vehicle. In the event the Henschel design came out clearly the best in the trials and was ordered into production as the Tiger tank.

The Porsche tank had an unconventional electro-mechanical drive. As 90 tanks had already been ordered, production was authorised but trouble with both the suspension and the drive brought work to a halt after only five had been built. Porsche was unable to solve the technical problems and in October 1942 the builders, Nibelungwerke, were ordered to cease production. Meanwhile Hitler had been pressing for a large turret for the VK4501 that would carry an 8.8cm KwK L/71. This had not been possible and in September 1942 it was decided that an alternative would be a heavy assault gun (Sturmgeschütz) with a frontal armour of 200mm (7.9in) and an 8.8cm gun. It was decided that the Porsche VK4501 vehicle could be converted to this since the parts were already in production. Even though there was a shortage of chassis parts and the mechanical problems had not been solved, Hitler ordered that the 90 VK4501(P) chassis should be converted to the new heavy assault gun requirement and be ready for the 1943 summer offensive on the Eastern Front. The conversion involved building a sloped superstructure to house a Pak 43/2 8.8cm gun on the rear of the superstructure so that the long gun would not project too far past the nose of the vehicle. The gun was the same as the standard Pak 43. The armour thickness was 200mm (7.9in) at the front, 95-85mm (3.7-3.4in) on the sloped sides of the superstructure, and 45mm (1.8in) on top. This made it an extremely well protected vehicle, but it suffered from unreliability and the lack of a close-in machine gun for self-protection.

Above: Interior views of the Pak 43/1 installed in the Nashorn.

| **Data:** | Sturmgeschütz mit 8.8cm Pak 43/2 Sd Kfz 184 Ferdinand/Elefant |
|---|---|
| **Crew:** | 6 |
| **Weight:** | 70,000kg (68.6 tons) |
| **Length:** | 8.14m (27ft 1.6in) |
| **Width:** | 3.38m (11ft 3.2in) |
| **Height:** | 2.97m (9ft 10.8in) |
| **Armament:** | 1 x 8.8cm Pak 43/2 L/71, 7.92 MG34 (later) |
| **Ammunition:** | 50 x 8.8cm (varied), 600 x 7.92mm |
| **Armour:** | 200mm (front), 95-85mm(super structure), 45mm (top), 80mm (side), 25-100mm (mantlet) |
| **Engine:** | Maybach HL120TRM |
| **Speed:** | 30km/hr (18.6mph) |
| **Range:** | 150km (93mph) |
| **Traverse:** | 12° left and right |
| **Elevation:** | 6° to 25° |

ABOVE: The Jagdpanther with Pak 43 L/71 gun and well-shaped, well-armoured superstructure was the best of the 8.8cm tank destroyers by far.

RIGHT: The mighty Elefant with Pak 43/2 8.8cm gun as first produced and in service for the July 1943 Battle of Kursk.

ABOVE RIGHT: The Tiger II 'King Tiger' was the biggest German AFV to mount the 8.8cm gun, having the KwK 43 L/71. This is the early production type with the Porsche-designed turret.

FAR RIGHT: The Elefants that survived Kursk were given cupolas and hull machine gun, as seen here, and went to the Italian front.

All 90 vehicles, named Ferdinand in honour of the designer, took part in the epic Kursk battle of July 1943 but they were too vulnerable and many were lost or immobilised. Those that survived were given cupolas and hull machine guns; renamed Elefant some saw service later in Italy.

### 8.8cm Pak 43 L/71 auf Pz Jagdpanther

By far the most successful application of the 8.8cm gun to a self-propelled chassis came with the adaptation of the famous Panther tank to create the celebrated tank destroyer popularly called the Jagdpanther (Hunting Panther). The Panther tank (PzKpfw V) was a very successful and reliable design, and a fast tank destroyer was evolved by making an armoured superstructure in place of the turret and utilising the Ausf (model) G chassis, the current production type in January 1944. The design was approved in December 1943 and the MIAG firm built 382 starting at the turn of the year. The 8.8cm Pak 43 was fitted in a heavy cast mantlet mounted in a steel ring welded to the front plate. Later vehicles had the mantlet bolted in place. The driver remained in the usual position and had a periscope next to the gun for forward vision. The crew all had individual roof periscopes. A *Nahverteidigungsgerät* (close defence weapon) was mounted in the roof and there was a machine gun mounted in the hull front. While the suspension was unchanged, a heavier transmission was fitted to compensate for the increased weight. Most of these vehicles were used on the Western Front in 1944-45. The Jagdpanther was a very efficient and reliable weapon. Firing APCBC shot, 184mm (7.25in) of homogeneous armour could be penetrated at a range of 457.2m (500yd), and 169mm (6.65in) at 914.4m (1,000yd).

| **Data:** | 8.8cm Pak 43 L/71 auf Pz Jag Panther (Jagdpanther) |
|---|---|
| **Crew:** | 5 |
| **Weight:** | 46,736kg (46 tons) |
| **Length:** | 9.9m (33ft) |
| **Width:** | 3.42m (11ft 4.8in) |
| **Height:** | 2.72m (9ft 0.8in) |
| **Armament:** | 1 x 8.8cm Pak 43/3 L/71<br>7.92mm MG34 |
| **Ammunition:** | 60 x 8.8cm rounds (varied)<br>600 x 7.92mm |
| **Traverse:** | 3° left and right |
| **Elevation:** | 14° to 80° |
| **Armour:** | 80mm (3.15in) front,50mm (2in) side, 100mm (4in) mantlet |
| **Engine:** | Maybach HL230P30 |
| **Speed:** | 46km/hr (28.5mph) |
| **Range:** | 160km (100 miles) |

BELOW: Most Tiger IIs had the squarer, bigger, Krupp/Henschel turret. Note the massive mantlet and the Zimmerit coating against magnetic mines. This was omitted from late production vehicles.

### PzKpfw VI Tiger Ausf B Sd Kfz 182

The most spectacular German vehicle to carry the 8.8cm gun was the famous Tiger II or Königstiger (King Tiger), heavily armoured and of formidable size and weight at over 69,000kg (68 tons). First proposals for this tank were made at a meeting in May 1941 when Hitler called for a heavily armed and armoured tank to 'spearhead' the attacks of the Panzer Divisions. For this purpose a 8.8cm gun was to be used. Development was marked by some conflict between the Henschel company and Dr Ferdinand Porsche who were both asked to submit design proposals just as they had for the VK4501 (Tiger I) programme. A Flak 41 installation was proposed by Porsche, but this presented difficulties and by February 1943 the view of *Waffenamt* (ordnance office) and Krupp had prevailed and the 8.8cm KwK 43 L/71 was finalised as the armament. This was shorter than the Pak 41 L/74 gun. Some Porsche turrets had been built intended to take the Flak 41, but they were adapted to take the KwK 43 for which Krupp had designed a turret that was used in most of the King Tigers produced. In early 1945 a proposal was made to develop a longer (L/105) version of the KwK 43 gun for this tank, but no progress was made on this before the war ended. A Panther II tank with a small turret also mounting the KwK 43 gun was in development in 1944-45 , but this never progressed far enough to reach production.

The large size of the Tiger II allowed a good stowage of ammunition, 80 rounds in all. In the standard Krupp/Henschel turret 22 rounds could be stowed, 11 each side, with 58 stowed in the hull. The Porsche turret, fitted to only a few vehicles, carried only 16 rounds. The first production order for the Tiger II was placed in October 1942 with delivery in September 1943. However, due to disruption from bombing and technical problems, the first vehicles were not in service until January 1944. Production continued until March 1945 and three prototypes plus 489 series vehicles were built. Over 600 others ordered were lost due to bomb damage at the factory, or delays caused by bomb damage. As a fighting machine the Tiger II had mixed fortunes. The Kwk 43 gun was a very accurate and powerful gun, and the ammunition supply for it ( usually 50 per cent HE and AP rounds) was adequate. The heavy armour protection made it difficult to knock out in tank-to-tank combat and armour penetration from the front was rarely, if ever, achieved. The heavy weight and great size, short engine life and frequent breakdowns all told against it and reduced its effectiveness.

| **Data:** | PzKpfw Tiger Ausf B |
|---|---|
| **Crew:** | 5 |
| **Combat weight:** | 69,800kg (68.7 tons) |
| **Overall length:** | 10.286m (34ft 3.44in) |
| **Hull length:** | 6.4m (21ft 4in) |
| **Width:** | 3.76m (12ft 6.2in) |
| **Engine:** | Maybach HL230P30 V12 700hp |
| **Max speed:** | 30km/hr (18.6mph) on road, 15-20km/hr (9.3-12.4mph) cross-country |
| **Range:** | 170km (106 miles) |
| **Armament:** | 1 x 8.8cm KwK 43 L/71, 2 x 7.92mm MG34 |
| **Ammunition:** | PzGr 39/40 (AP), PzGr 40/43 (AP), SprGr 43(HE), HlGr 39 (HC) |
| **Stowage:** | 80-84 rounds |
| **Elevation:** | -8° to +15° |
| **Traverse:** | 360° |
| **Max range:** | 10,000m (10,900yd) with SprGr 40/43 |
| **Muzzle velocity:** | 1,000m/sec (3,280ft/sec) with PzGr. 39/43 |

Special measures were needed to overcome the weights involved. The Porsche turret vehicles had a one-piece Kwk 43 barrel (its trunnion positions were further forward) and the Krupp/Henschel turret had a two-piece gun. A hydro-pneumatic cylinder was fitted between the recoil cylinders and the mounting to counter the barrel weight when elevating the gun. To speed up traversing, a hydraulic motor was fitted to the turret drive but manual traverse was also possible. A standard monocular sight was fitted, TZF 9.

### 8.8cm Pak 43/3 auf Panzerjäger Tiger Ausf B

The Jagdtiger, a tank destroyer with fixed armoured superstructure was a development of the Tiger Ausf B chassis. The designed armament for this very heavy vehicle was the 12.8cm Pak 44 or 80 L/55. However, there was a shortfall in supply of the l2.8cm gun, mainly due to the bombing of Krupp's works where the guns were made, so on a few vehicles an 8.8cm Pak 43/3 was substituted.

## Miscellaneous prototypes and projects

There were a number of other adaptations of the 8.8cm gun to self-propelled chassis, some of them experimental and some only mock-ups. Some were in the Waffenträger (weapon carrier) programme. None reached production status. Among these projects were a Flak 41 on a Panther tank chassis (project model only), a KwK 43 on a Sd Kfz 251 chassis (running prototype built), Pak 43/3 on a PzKpw 38(t) chassis, Pak 43/3 on a Krupp/Steyr Sfl 38(t) chassis, and a Pak 43/3 on a Sfl 38(t) chassis. These last three were variants of Waffenträger, each featuring different styles of mount for the 8.8cm Pak 43/3 on modified chassis of the ex-Czech PzKpfw 38(t) tank

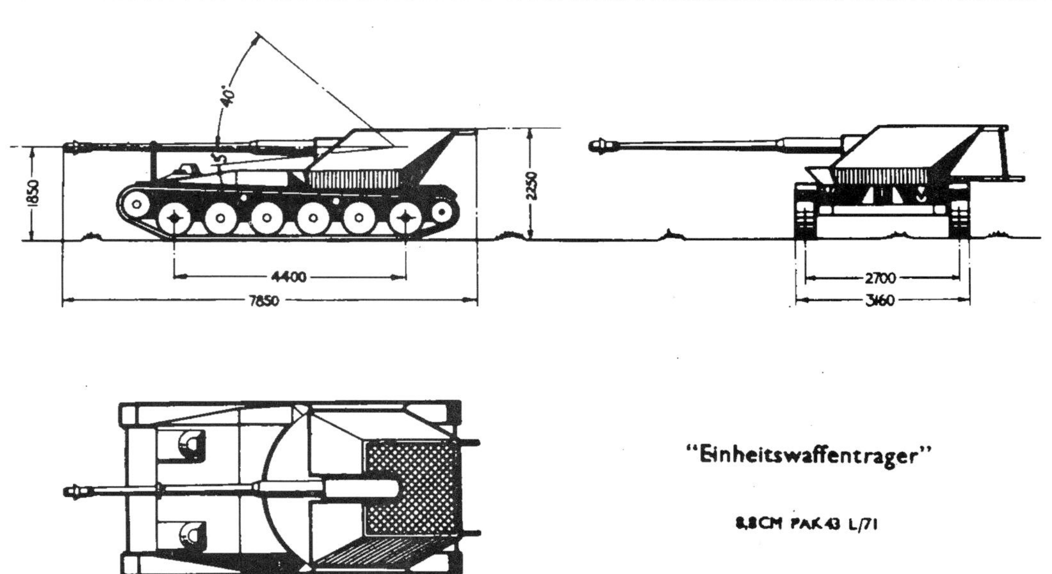


TOP: Proposed fitting of the 8.8cm Pak 43 on the Sfl chassis, named Grille 10.

ABOVE: Krupp-Steyr prototype 8.8cm Pak 43/3 Waffenträger.

RIGHT: Waffenträger project for a Pak 43 L/71 on a modified, lengthened PzKpw 38(t) chassis.

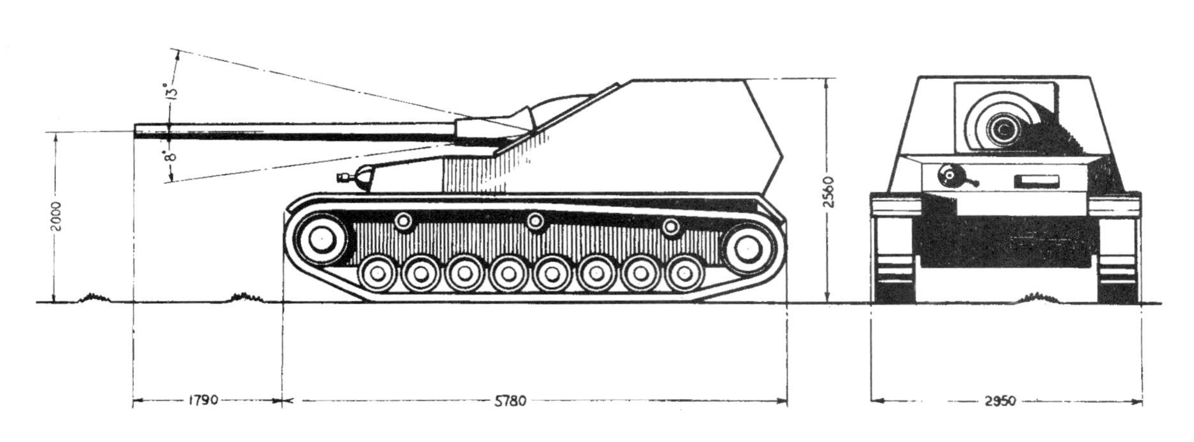


LEFT: Plans for a proposed Panzerjäger IV with an 8.8cm Pak 43/3, on a PzKpfw IV tank chassis.

BELOW LEFT: Rheinmetall-Borsig prototype for 8.8cm Pak 43 on a standard Waffenträger chassis.

Below: Ardelt prototype for an 8.8cm Pak 43/2 Waffeträger.

RIGHT: Mock-up for a proposed Flakpanther featuring a Flak 41 on a Panther tank chassis. It was never built.

FAR RIGHT: Second prototype for the Ardelt 8.8cm Pak 43/3 Waffentrager.

BELOW RIGHT:: Versuchsflakwagen fitted with a Flak 37: it saw limited field trials in this form in Italy.

## Railway gun mounts

A further category of mobile mounting featuring the 8.8cm gun was the so-called 'flak train' (*Eisenbahnflak*) which was a measure to combat the heavy Allied bombing offensive over Germany. Whole trains of anti-aircraft guns could be moved from area to area wherever bombing raids were being concentrated, or perceived to be coming.

Numerous improvised mounts were used often featuring standard railflat cars or low side cars with Flak 18, 36, 37, or even 41 guns fixed on. In some cases wood baulks were used on the sides for crew protection. Some long two truck flat cars carried two 8.8cm guns with an ammunition stowage locker in between. Passenger cars were also used for personnel accommodation and ammunition carriage. Apart from the many improvised rail flak cars, there was a standardised conversion, the Geschützwagen III (Eisb) schwere Flak. This was a four-axle flat car with side rails, modified by removal of the brakehouse and the uprights and with heating pipes added. This car could carry the 8.8cm Flak 18, 36, or 37 or the 10.5cm Flak 38 and 39. This car had drop sides which provided a working platform for serving the gun, ammunition lockers at each end, and screw jacks to stabilise the car when firing. Overall length 15.8m (52ft 8in), weight loaded 35,000kg (34.5 tons), weight with gun and ammunition on board 45,800kg (45 tons).

BELOW: *Geschützwagen III (Eisb) Schwere Flak* with Flak 18 or 36 mounted, and showing sides dropped for action. Note ammunition stowage lockers. This was the standard railway mounting, though extemporised ones were also used.

# Combat & Deployment

ABOVE: The business end of the Pak 43/1 installed on Nashorn.

ALL THE EVIDENCE suggests that the highly effective use of the 88 in the anti-tank role in World War 2 had its groundings in World War 1. As Rommel and other German senior officers had all served in the previous war and through the interwar years they were clearly versed in what had been done before. Rommel's deployment of his Flak 88 guns as anti-tank guns at the Battle of Arras in May 1940, and again at Halfaya Pass in 1941 was probably not the sudden inspiration that it appears to be. In fact the Germans first used their anti-aircraft guns in the anti-tank role with some success in World War 1 and, furthermore, had a round suited to tank penetration.

**Developments in World War 1**

As noted earlier, a 1909 infantry versus balloon trial established that fire heavier than rifle calibre was necessary to bring down a balloon, but at the time the German War Ministry took no immediate action to achieve this. One reason for this seems to have been doubts about the means of deploying the guns, largely because the Inspector General of Military Vehicles considered that the new motorised BAKs were not reliable enough mechanically. This caused the Krupp company to offer a much improved equipment in 1911. The Daimler carriage had its track width increased (but retained the same wheelbase as before) and was given a beefed up transmission with two final drive shafts giving eight forward and two reverse gears. An improved version of the 7.7cm L/27 BAK was fitted. This still failed to impress the Inspectorate, however, and they stated officially that the army had no further interest in motorised guns. This in turn stirred the Inspectorate of Artillery which was alarmed at the prospect of the future for it realised that there was not only an increasing threat from the air, but that motor tractors would be needed to haul field artillery in any future major conflict.

The Inspectorate of Artillery therefore persuaded the Inspectorate of Military Vehicles to authorise further trials to investigate 'devices for driving vehicles on dirt and gravel tracks'. Various road vehicles and motor tractors were therefore tested in the 1911-13 period. One result of this was authority from the War Ministry to order an improved 1912 BAK with both Krupp (with a Daimler truck) and Rheinmetall (with its own Erhardt truck) involved in meeting the requirement. A four-wheel drive 70hp truck chassis was called for—advanced thinking for the time —and sand rims were to be fitted on the wheels to assist running over soft off-road surfaces. Later the power requirement was increased to 80hp. To improve low speed traction on its design, Erhardt introduced two extra reduction gears, one between first and second gear and the other between third and fourth gears. Daimler tackled the same problem by introducing an intermediate geared drive. An irritation here was that the vehicle had to stop to change to the intermediate gear before going off-road.

Trials with these prototypes, and other vehicles, had a useful side effect for it made both the Artillery and Military Vehicles departments realise the potential value of four-wheel drive. However, they were cautious for several reasons. First there was the problem of the extra cost of four-wheel drive vehicles. This was allied to rapidly advancing technology which might render vehicles outdated very quickly. Finally there was the question of needs; aside from the artillery who else would use them? Hence it was decided on very limited procurement only for the time being.

In 1913 some production orders for

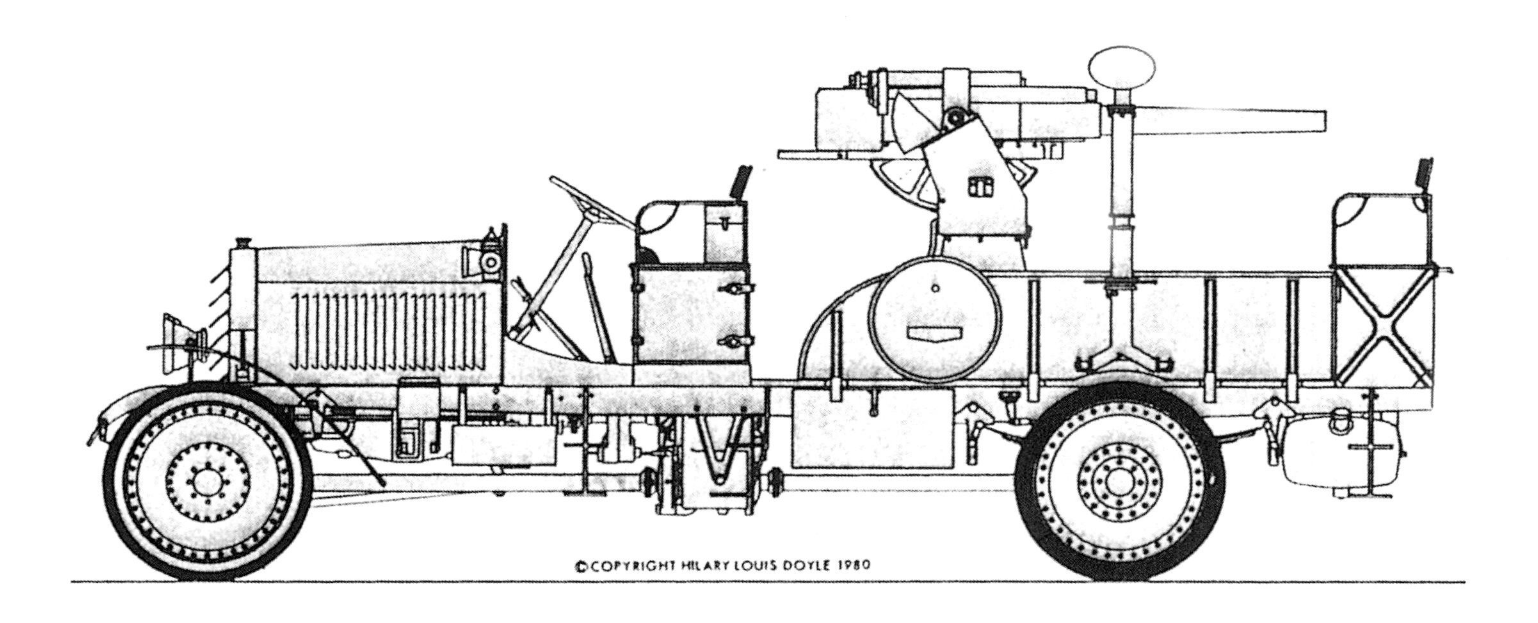


ABOVE: The Rheinmetall-Erhardt improved model of the 7.7cm BAK showing four-wheel drive arrangement. *Drawing by Hilary Doyle*

motorised BAKs were finally forthcoming. From Erhardt came a batch of 7.7cm L/27 BAKs, designated Model 1914, and used for military trials. This vehicle had a four-cylinder gasoline engine on a steel sheet chassis frame. The gearbox was suspended centrally below the chassis giving four forward and one reverse gear. There was a third drive shaft below the ordinary drive shafts to give extra power to all four wheels. The Krupp company, which worked with Daimler on the automotive aspects, was asked to provide two BAK vehicles, described as 'platform trucks' presumably because the dropsides formed a platform for the gun crew to work the equipment. The Daimler truck used met the new specifications, having a 70hp motor, increased track width (1.53/1.68m—5ft/5ft 6in), four-wheel drive, and a 7.7cm L/27 BAK Model 1914. The complete equipment weighed 7,035kg (7 tons) and the detachment was 10 men, including commander and vehicle driver.

The Krupp/Daimler and early Erhardt deliveries seem to be among the six only motorised BAKs the German Army is recorded as having in service when war broke out on 4 August l914, but the coming of war gave impetus to the whole programme as might be expected. Besides the balance of the order for Erhardt motorised BAKs , further orders were placed. Over 100 new vehicles were ordered from both companies, 57 from Krupp and probably the same number from Erhardt. Both vehicles had a 7.7cm L/27 gun though they were different designs—a Krupp gun on the Daimler vehicle and a Rheinmetall gun on the Erhardt vehicle. As aircraft became the main aerial enemy the designation of the gun and equipment was soon changed to 7.7cm leichte Kraftwagen-Flak L/27 1914. For short they were often referred to as Kraftwagen 1914 (motor truck 1914). In 1916 the designation was changed to *Kraftwagen-Flug-Abwehrkanone* (motor air defence gun) and because of their mobility they continued through the war as a most important piece of field ordnance. Later in the war they were called K-Flak or Kraftwagen Flak.

### The first tank destroyer?

While the place of the Kraftwagen 1914 is firmly established as a key stage in the development of anti-aircraft equipment, it is a much less well-known fact that it may well have been the first self-propelled tank destroyer, pre-dating what most take as a World War 2 development by some 25 years. As soon as armoured vehicles and emplacements were perceived as a threat the German Ordnance Department developed an armour-piercing round for the 7.7cm gun, designated KGr.15P. This shell was able to penetrate 30mm (1.18in) of armour at 3,000m (3,280yd).The K-Flaks were probably first used in the anti-tank role in Champagne against the first French tank attack of World War 1 in the spring of 1917. According to German military histories the next deployment of the vehicles was at the Battle of Cambrai in November 1917 where a battery of K-Wagens was responsible for the carnage in the Flesquieres sector where 16 British tanks were destroyed by gun fire and the advance was effectively stopped in that sector. The Germans claimed that 50 British tanks at Cambrai were knocked out by K-Wagens firing AP rounds. This may well be true, for British tank losses to gunfire were heavy—65 in all on the first day of the battle. This was actually fewer, however, than were lost by mechanical defects or stranding in ditches, etc.

| **Data:** | 7.7cm I Kraftwagen-Flak L/27 M1914 |
|---|---|
| **Calibre:** | 7.7cm |
| **Length of barrel:** | 2.080m (6ft 11.2in) |
| **Breech mechanism:** | Automatic sliding wedge |
| **Gun mount:** | Pedestal |
| **Elevation:** | -5° to 70° |
| **Traverse:** | 360° |
| **Ammunition:** | KGr.15 (Flak), KGr.15P (AP) |
| **Shell weight:** | 6.85kg (15lb) |
| **Maximum height:** | 4.25m (14ft 2in) |
| **Maximum range:** | 7,800m (8,500yd) |
| **Muzzle velocity:** | 465m/sec (1,525ft/sec) |
| **Rate of fire:** | 20-25rpm |
| **Weight of gun:** | 1,082kg (2,380lb) Krupp, 990kg (2,178lb) Rheinmetall |
| **Weight of vehicle:** | 3,125kg (6,875lb) Krupp/Daimler, 7,030kg (15,466lb) Rheinmetall/Erhardt |

The K-Wagens were used in subsequent actions against tanks and in this role they were known as *Tankzug auf Kraftwagen.* Interestingly enough the British commanders did not seem to realise exactly why they lost so many tanks to gunfire, or if they did they were not letting on. In a curious forerunner of reaction to the 1941 Halfaya

RIGHT: The production K-Wagen—7.7cm leichte Kw-Flak L/27 M1914. Note ammunition locker and sand rims on wheels. This is the Rheinmetall model.

Pass debacle, where a 'wonder weapon' was attributed to the British tank losses, in his report on the heavy tank losses at Flesquieres, Field Marshal Haig attributes the German success to a determined German artillery officer who 'served a field gun single handed until killed at his gun'. This seems to have been only partly true, for though one story tells of an Unteroffizier named Kruger who hit five British tanks as they crossed his line of fire, there was certainly no superhuman act of valour to account for the British losses. German accounts merely told of several gun batteries engaged in direct fire who had had anti-tank training, but even the German stories play down events and the significance of this early used of self-propelled tank destroyers is not widely known.

**The original 8.8cm requirement**

Krupp and Rheinmetall first suggested that 8.8cm and 10.5cm would be the best calibres for the next generation of anti-aircraft guns in a submission to the German War Ministry in July 1915. This led to a meeting in early 1916 between Krupp, Rheinmetall, the War Ministry, the Inspector of Flakartillerie and the Artillery Test Commission. The desirable characteristics and specification for the new guns was drawn up then and these included features that were to endure through much of the future developments of the 8.8cm gun in particular. The requirements included:

- Gun mounting by pedestal on a trailer with reversible side stabilisers and supports with spring mounts.
- Elevation from +20° to +70°.
- Pneumatic recuperator above, barrel brake underneath the barrel, rearward set trunnions, variable barrel recoil and spring compensator.
- Barrel of constant rifling with end rifling as in the 8.8cm Marine-Flak L/45 (naval) anti-aircraft gun already in service. Semi-automatic breech.
- Sighting devices to include telescopic sight with deflection calculator calibration and independent height/altitude estimation.
- Ballistic performance (8.8cm) of 750m/sec (2,460ft/sec) muzzle velocity.
- Shell weight 9.6kg (21.2lb), fuse S29.

It was suggested the existing 8.8cm Schiffskanone L/45 in Mittelpivotlafette M1913 should serve as the pattern. This naval gun had already been used successfully on its pivoted mount in the AA role.

The Krupp and Rheinmetall guns, 8.8cm K-Zugflak L/45, that resulted from this specification proved very successful (see Early Development chapter). Each gun was supplied with a four-wheel drive towing tractor, Daimler for the Krupp and Erhardt for the Rheinmetall. The truck transported the gun detachment and carried ammunition. The towing speed was low, though, at 12km/hr (7.46mph), although it seemed less so at the time when horses were even slower and vehicle speeds were low in general.

There were interesting variations between the two makes of gun. The Krupp gun had a one-piece (monobloc) barrel with screwed on breech and counter-ring. It was semi-automatic with right and left cartridge discharge. The traverse sighting mechanism had worm gearing and the elevation sighting mechanism had planetary gearing. The Rheinmetall gun had a jacketed tube with a covering ring, semi-automatic breech, elevating mechanism with toothed quadrant movement, and traversing mechanism with worm gearing.

The Rheinmetall Model 1917 version of the gun introduced important improvements. The elevating and traverse mechanisms were altered so that the gun had a line of sight independent of height and traverse. This allowed the setting of the angle of sight/tangent elevation and the deflection lead angle without altering the line of sight, a key need with ever-increasing aircraft speeds.

The other key development was the

linked predictor (*Kommandoeinstell-Vorrichtungen*) which was joined to the right side of the gun. This gave a graduated disc and movement shaft with pointer marks for traverse, and a graduated drum (0-30°), a meter division drum in three stages, and a graduated regulator (60 divisions) for elevation (height). The sighting equipment on the left of the gun consisted of a linked deflection calculator, a ground angle indicator, a Flakzielfernrohr 18-telescopic sight and carrier, and the linkages. This revolutionary device allowed predicted deflection to be continuously transmitted to the sights with no need to consult tables or conversion charts.

The 8cm and 10.5cm K-Flak guns produced at the same time as the 8.8cm guns followed the same design features and differed only in size and barrel recoil. The 8.8cm guns were produced in by far the greatest numbers during World War 1: figures show 160 8.8cm guns delivered while the 8cm and 10.5cm guns were only produced in small numbers. These guns appear to have been only used in the AA role.

**Deploying the 88**

Notwithstanding the interesting development of anti-tank and anti-aircraft guns in World War 1 and the useful experience that must have been gained, the more dramatic use of the 88 came in World War 2. Its most spectacular exploits were in the anti-tank role in the field, but it must not be forgotten that its prime role was anti-aircraft work and for every gun delivered on a wheeled carriage for the field forces there were three delivered without carriages and in ordinary pedestal form for emplacement in fixed AA defence positions. Guns produced in pedestal mount form only were designated Flak 18/2, Flak 36/2, Flak 37/2 and Flak 41/2. They were either bolted down or bedded in concrete in fixed AA emplacements. A special low loader trailer was produced, Sonderhanger 205, to carry the pedestal guns from site to site, but this was not in use in large numbers so a fair number of emplaced 88 were abandoned in retreats in the closing months of the war as there was no means of carrying the guns away. Anti-aircraft batteries were a

RIGHT: Highly successful—they nearly forestalled the Omaha beach landing—were well emplaced 88s like these Flak 18s in concrete protected bays as part of the Atlantic Wall.

LEFT: A Flak 18 in a coastal defence emplacement firing on exercises.

BELOW: A Flak 36/2 fixed mount on a concrete base, typical of AA defence positions.

Luftwaffe responsibility and the 88 in its various forms could be found all over occupied Europe and in the Reich itself in fixed AA defences. Some old Flak 18s and 36s in pedestal mounts found their way into coastal batteries of the Atlantic Wall along with the assortment of other miscellaneous ordnance used in this way.

AA batteries of all calibres were also stationed in critical parts of the Atlantic Wall—for example at estuary entrances—and these included 88s in some cases. One of the most notable—and successful—examples of the use of the 88 in the coast defence role was at Omaha beach on D-Day, 6 June 1944. Here the American forces had an extremely tough time getting ashore and suffered huge losses, only just managing to secure a beach-head after a day of bloody fighting. They had the misfortune to be up against the best German infantry division guarding the Normandy coast, but the nature of the low ground inland from Omaha beach allowed the 88s used there in protected emplacements to both enfilade the beach and fire out to sea. The first gun to open fire against the approaching invasion forces was an 88 that fired 10 rounds in rapid succession against the veteran 32-year old battleship USS *Arkansas,* lead ship of the gun line. This

happened at 05.30 just as soon as it was light enough to see. All the shots fell short but it was sufficient for the *Arkansas'* gunnery controllers to get a range on the position and take the intrusive 88 out with one well-judged 12-inch round.

This was one of the few successes of the day. From H-Hour at 06.30 until mid-afternoon, the US Army forces struggled to get ashore, hampered by incessant mortar, machine gun, and 88 fire. The first two LCTs to hit the beach were holed and set afire by 88s and the third only survived because it was shielded by the other two. When the M4 Sherman tanks left these LCTs the 88s switched fire to them and one was on fire before it cleared the water. This set the pattern for succeeding waves of incoming LCTs and LCIs, the 88s harassing the landing craft, then enfilading the troops and vehicles as they tried to land. At least one LCT was hit and immobilised half a mile out to sea while evacuating wounded. So many LCTs and LCIs were shot up and stranded on or near the beaches that they obstructed the next landings. Several LCIs were sunk and the soldiers they were carrying had to swim ashore. Because insufficient tanks and artillery pieces could be landed there was a lack of heavy firepower on the beaches. At about noon the destroyers in the Omaha Beach fire line—eight American, three British—were sent in dangerously close to the shore line, just 730m (800yd) offshore with the task of taking on and eliminating the 88s and machine gun posts. At times these ships had just inches of water below the keels. This close-in fire support played a big part in saving the day. Further out the old battleships *Texas* and *Arkansas* fired 771 14-inch and 12-inch rounds, some inland to disrupt German reinforcements, but others onto the beach itself and into the gun positions on Pointe-du-Hoc. A salvo from the *Texas* was famously used to blast German positions out of the main beach exit later in the afternoon.

RIGHT: Luftwaffe crewmen man a Flak 18 during a night raid over the Reich. These are starshell rounds being loaded.

Luftwaffe anti-aircraft organisation had a good reputation for efficiency with well co-ordinated predictors and radar installations, and radar-controlled searchlights. Though the 10.5cm AA gun became more important, the 88 was in service in much greater numbers. The Flak 41 was increasingly important and more were allocated to Luftwaffe Flakartillerie units than went to army field units, even though the Flak 41 had a dual use role. But its expansion and extraction problems made it more of a liability in the field. It was only in the last year of the war that percussion fuses were available for AA use, and the normal way of bringing down an aircraft was by air burst close to the aircraft, using a time fuse. Allied bombers could be—and were—shot down like this. Again the Flak 41 was a better weapon for AA work as it had its fuse-setting machine alongside the loading tray ensuring less delay in getting a time fused round away. Luftwaffe Flak organisation was very varied and could be changed or adapted to suit requirements. However, while corps, brigade and district commands could differ, the basic Flakregiment usually had a staff/command section and four *Abteilungen* (battalions). Abteilung I and II were gun units, Abteilung III was the searchlight unit and Abteilung IV was a training or support unit. Each Abteilung was divided into three batteries. A heavy Flakbatterie would have four 88s and two 20mm guns when part of a field division. The 20mm guns were to protect the 88s and were sometimes vehicle-mounted. The battery also had a predictor. In home defence, batteries in fixed locations would not have the 20mm AA guns and would, instead, have six 88s. The Flakbatterie was divided into three platoons (*Flakzug*) each with two guns.

Army organisation of the 88 varied greatly through the war and depending on the theatre of war. Also different types of division (infantry, panzer, etc) had different allocations of weapons. An army AA battalion (*Heeresflakabteilung*) was fully motorised and normally had two or three batteries each with four 8.8cm guns (and a predictor), plus one or two light batteries each with 12 x 20mm guns or 9 x 37mm guns. The 8.8cm guns were fitted for both AA and anti-tank operation with the necessary ammunition mix. Each front line division usually had an AA battalion on its strength, so the divisional commander

ABOVE: Flak 18 with shield in the AA role, manned by the Luftwaffe.

LEFT: The classic 88 deployment as an anti-tank gun in direct fire. This Flak 36 is well camouflaged in an ambush position, probably in Normandy in 1944. Firing from its carriage it will be able to make a fast getaway to avoid counter-battery fire.

BELOW: A Flak 18 of a divisional Flakartillerie regiment, manned by the Luftwaffe, is quickly emplaced to be used in direct fire. Note the wood/wicker baskets for ammunition supply in the foreground.

RIGHT: Flak 18, without shield, manned by the Luftwaffe.

could call on either 8 or 12 x 8.8cm guns for anti-tank work or indirect fire tasks. The towing vehicle for the 8.8cm was the Sd Kfz 7 half-track which had seating for the gun crew and an ammunition locker. However, sometimes other towing vehicles were used. When the Pak 43 and Pak 43/41 guns entered service they were allocated to Festung-Pak companies within the divisional anti-tank battalion. A company normally had nine guns.

While the Flak 18 and 36 were highly successful in the anti-tank role, their use was not without hazard to the users. They stood high on the ground and were hard to conceal if time was not available to set up a prepared defensive position. Both the gun and its crew were extremely vulnerable to counter-battery fire and to ground-attack aircraft in the latter part of the war. The front shield on the mounting offered only minimal protection to the crew, and none at all to the ammunition supply numbers. On the other hand the gun could be brought quickly into action on its carriage which was very useful in a war of movement. The later type of limber, Sondergerät 202, added to this facility since it allowed the gun to be towed with the barrel facing to the rear so it could even be fired in emergency while still hooked to the towing vehicle. With the earlier limber the gun was towed with the barrel facing forward, so it was necessary for the towing vehicle to pull clear before the gun could go into action. However, with a well trained crew the 88 was always a very formidable weapon in either the anti-aircraft or the anti-tank role.

# Appendices

ABOVE: Flak 36/43s with crews under training.

## 1 Description of the 8.8cm Flak 36

**Barrel**
Detachable breech ring with a half-length outer tube, a half-length inner lock tube, and a three-piece liner in three sections to facilitate barrel-wear replacement. Front and centre sections keyed to align rifling. Locking collar and locking ring prevent movement.

**Breech**
Semi-automatic horizontal sliding block. A round of ammunition when pushed into the breech trips extractors which allows the breech block to close under pressure from an actuating spring.

**Firing mechanism**
By percussion mechanism composed of firing spring, firing pin and firing pin holder. Cocking achieved automatically during opening of breech. Automatic firing as breech closes unless the cocking lever is moved to the 'safe' position.

**Recuperator**
Set above the barrel, secured to the cradle; independent hydro-pneumatic system.

**Recoil cylinder**
Set below the gun inside the cradle, remained stationary in recoil.

**Carriage**
Box-section construction, welded and riveted, designed to form a carriage with bogies (or limbers) each carrying a winch to raise and lower the carriage. Side outriggers hinged to main carriage, with levelling jacks. Electrical wiring inside the box-section arms. Pedestal bolted to bottom carriage, with data transmission box at rear of bottom carriage. Top carriage levelled by handwheels, with level indicator on pedestal. Pedestal features traversing ring and levelling universal. Azimuth scale provided for orientation.

**Cradle**
Rectangular trough type, welded and riveted. Slides of the cradle supported and guided the gun in recoil and counter-recoil.

**Equibrilator**
Two springs to balance the muzzle weight, suspended by trunnions and enclosed in telescoping housings.

**Traverse**
Handwheel on right of mount, high or low speed with selector lever on wheel. All round traverse. Belleville spring stops movement after two complete turns one way to prevent tangling of data transmission cables. Azimuth data transmission indicator geared directly to the rack on the traversing ring.

**Elevation**
Handwheel on right of carriage. Motion transmitted through gears to the elevating pinion. High or low speed with selector handle on wheel. Clutch provided to disengage the elevating mechanism to prevent damage when towing.

**Bogies (or limbers)**
Single wheels, 7-leaf transverse spring on leading bogie. Twin wheels, 11-leaf transverse spring on rear bogie. Cast spoked wheels. Air brakes to all wheels. Emergency hand brake on rear bogie. Adjustable height drawbar. Bogies are fitted to take a tube joining bar to make an improvised trailer when removed from the

carriage. Hand brakes set on rear bogie when firing from the carriage. Hand winch with chains used to lower and raise carriage to bogies.

### Rammer

Provided to facilitate loading with gun at high elevation. Mounted on left top of cradle and actuated by hydro-pneumatic cylinder. Rammer arm protected by a folding guard.

## 2 Fire control equipment

Sighting and fire control for the Flak 18 and 36 varied depending on type of engagement, ie, direct fire, indirect fire or AA fire.

### Sights

For direct fire the telescopic sight ZF20E was used to lay in azimuth and elevation (ZF=*Zielfernrohr*=telescopic sight). When used against moving land, sea, or air targets in conjunction with the predictor/director a panoramic telescope RblF32 was placed in a holder on top of the recuperator to establish the initial orientation of the gun with the director. (RblF—*Rundblickfernrohr*—panoramic telescope sight.) The ZF20E had x4 magnification with field of view of 17° 30′. The range quadrant and elevation mechanism moved the entire sight. Deflection could be set into the sight by rotation of a deflection knob graduated 250 mils to right and left. Sight was removable and was clamped in place. The earlier ZF20 sight, also used, lacked a range drum.

### Fire control

The 88's primary equipment was the Kommandogerät 36 used in conjunction with the azimuth and elevation indicators and the fuse setter on the mounting. It was linked to the mount by cables. The Kommandogerät 36 transmitted data electrically to the azimuth and elevation indicators and the fuse setter, the operators on the gun matching the lights (or following pointers on later models). The KdoGr 36 featured a stereoscopic range finder adapted for height finding, with a range scale of 500m (550yd) to 50,000m (55,000yd). A device to judge approximate height was fitted to the rangefinder end to assist quick acquisition of the target. Two tracking telescopes were fitted on the rangefinder tube.

LEFT: A Flak 36 in action against Allied tanks in the Western Desert, 1942, typically well dug in as part of a defensive position. In the foreground the battery commander is using his telescope to observe fall of shot.

RIGHT: Rear and Front (Far right) panels of the auxiliary director.

BELOW: The Kommandogerät 36 fire control director/predictor on its trailer with the rangefinder in its carrying box alongside.

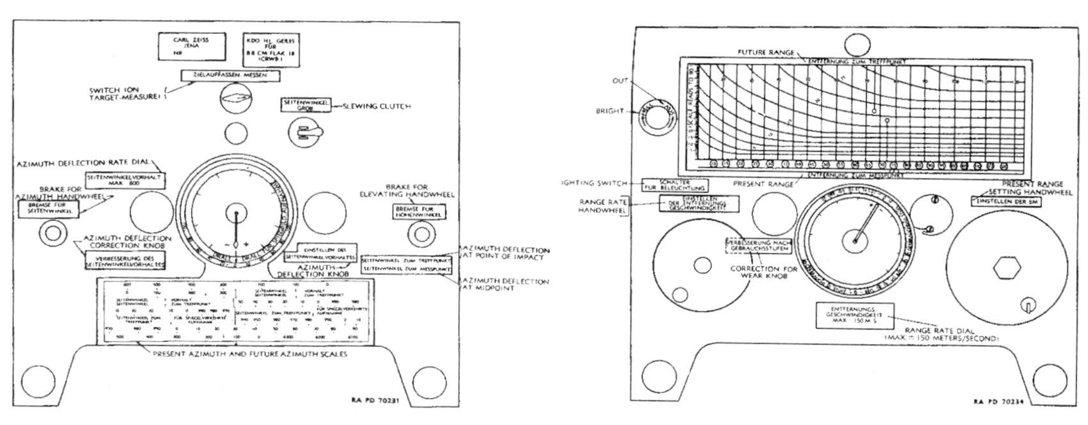


*Auxiliary Director 35 (Kdo. Hi. Gr. 35) — Rear Panel*

*Auxiliary Director 35 (Kdo. Hi. Gr. 35) — Front Panel*

The director predicted and transmitted to the battery all the following: quadrant elevation, future azimuth, time of flight of round expressed in fuse units, present angular height, present azimuth, present slant range. Data transmission was continuous. This equipment required a crew of 11 to track the target, take readings, and feed in information using handwheels, cursors and clockwork and electrical mechanisms incorporated in the equipment. The KdoGr 36 was supplied one per gun battery. Cables and a distribution box were included and the rangefinder was carried in a separate box and attached when the equipment was emplaced. The KdoGr 36 was carried on a short cruciform carriage similar in construction to that of the gun and used the same sort of Sondergerät 201 or 202 bogies. It had its own towing vehicle.

An alternative auxiliary director, Kommandohilfsgerät 35, was also available. This was a portable unit, much simplified from the KdoGR 36. It could be carried from a bar and was towed on a two-wheel trailer. It did not transmit data by cables. This was done by relaying information to the mounting by a field telephone. The instrument stood on a stand which had three levelling screws. Limit of range was 12,000m (39,360ft) and height limit 10,000m (32,800ft). The equipment needed a crew of nine providing, by tracking or setting, the following information: azimuth, elevation, range, azimuth rate, elevation rate, range rate, future azimuth, future elevation, fuse setting. The equipment had two tracking telescopes and operation was entirely mechanical, the only electric power being for the telephone.

Indicators on gun carriage: Azimuth and elevation indicator dials were fitted on the right side of the gun pedestal. On the Flak 18 and 36 pointers were moved to match rings of lights. On later models the same data was obtained by matching pointers. With the light system the gun was operated in azimuth and elevation by keeping the lights blacked out by the pointers. On later models it was operated by matching the pointers. The fuse setter was on the left side of the top carriage, manually operated, but fuse data was also transmitted via a system of lights followed by pointers. The pointers were geared to the fuse dial and the setting ring on the fuse setter.

When using the Kommandohilfsgerät 35 the data telephoned to the mounting was, instead, followed on a scale marked on the fuse setter. A setting crank on the front of the fuse setter controlled keys to cut the fuses. Two fuses could be cut at the same time. For connection of the data transmission cables from the KdoGr 36 a special 104 pin terminal was fitted on the end of the rear trail.

**Other equipment**
Also provided was an aiming circle, used for setting up battery, spotting, and determining azimuth angles; a portable Rangefinder 34, operated on the shoulders and transported in a carrying case; surveying and plotting rules; and a battery commander's telescope featuring a x10 binocular and used for orienting and observing fall of shot.

## 3 Specification for 8.8cm Flak 18

This information is reproduced from the US Army intelligence report on the equipment.

**A. GUN**
**Type:** Tube and loose 3-section liner
**Total weight**: 1,336.7kg (2,947lb)
**Weight of removable components:**
Breech ring—229.28kg (505.5lb)
Outer tube—356kg (785lb)
Inner tube—365.36kg (805.5lb)
Liner (muzzle section)—272.1kg (600lb)
Liner (centre section)—90.2kg (199lb)
Liner (breech section)—26.3kg (58lb)
Retaining rings—15.4kg (34lb)
**Overall length:**
Tube—4.7m (15ft 5in)
Gun and tube—4.94m (16ft 2.1in)
**Length in calibers:** 56
**Distance from centre line of trunnions to breech face:** 0.17m (6.5in)
**Travel of projectile in bore:** 4m (134ft 1.4in)
**Vol of chamber:** 3,703.4 cu cm (226cu in)
**Rated max powder pressure**:
c212,902.8kg/sq cm (33,000lb/sq in)
**Muzzle velocity:** 819m (2,690ft)/sec
**Maximum range:**
Horizontal—14,813.28m (16,200yd)
Vertical—11,887m (39,000ft)
Maximum effective ceiling at 70° elevation—7,620m (25,000ft)
**Rifling:**
Length—400cm (157.4in)
Direction—Right hand
Twist—Inc. 1 turn in 45 Calibers to 1 turn in 30 calibers
Number of grooves—32
Depth of grooves—1mm (0.0394in)
Width of grooves—5mm (0.1969in)
Width of lands—3mm (0.1181in)
**Type of breech mechanism:** Semi-automatic horizontal sliding block
**Rate of fire:**
practical at mechanised target—5 round/min
practical at aerial target—20 round/min

**B. RECOIL MECHANISM**
**Type:** Independent liquid and hydropneumatic
**Total weight**: 237.68kg (524lb)
**Weight of recuperator cylinder:** 129.2kg (285lb)
**Weight of recoil cylinder:** 108.4kg (239lb)
**Weight of recoiling parts in recoil mechanism:** 49.2kg (108.5lb)
**Total weight of recoiling parts (with gun and tube):** 1,432.9kg (3,159lb)
**Type of recoil:** Control rod type with secondary control rod type regulating counter-recoil
**Normal recoil:**
0° elevation—105cm (41.5in)
25° elevation—85cm (33.46in)
Maximum elevation—70cm (27.75in)
**Capacity of recoil cylinder:** 9.46l (2.5gal)
**Capacity of recuperator cylinder:** 17l (4.5gal)

**C. MOUNT**
**Weight (less cannon and recoil mechanism):** 2,561kg (8,404lb)
**Maximum elevation:** 85°
**Maximum depression:** -3°
**Traverse:** 360°
**Loading angles:** All angles
**Height of trunnion above ground (firing position):** 1.5m (5.2ft)

**Height of working platform (firing):** 24.3cm (0.8ft)
**Height of trunnion above working platform:** 1.34m (4.4ft)
**Leveling:** Pivots located 45° from either side of centre line of front outrigger (total of 9° each)
**Number of turns of handwheel to elevate from 0 to 85°:**
High gear—42.5
Low gear—85
**Elevation for one turn of elevating handwheel:**
High gear—2° (35.4mils)
Low gear—1° (17.7mils)
**Number of turns of handwheel to traverse 360°:**
High gear—100
Low gear—200
**Traverse for one turn of handwheel:**
High gear—3.6° (63.8mils)
Low gear—1.8° (31.9mils)
**Effort required at elevating handwheel** (in-lb)

| To elevate | High gear | Low gear |
|---|---|---|
| 0° | 55 | 110 |
| 20° | 110 | 160 |
| 40° | 192 | 110 |
| 60° | 214 | 55 |
| 80° | 209 | 50 |

| To depress | High gear | Low gear |
|---|---|---|
| 0° | 275 | 220 |
| 20° | 193 | 28 |
| 40° | 138 | 50 |
| 60° | 110 | 77 |
| 80° | 165 | 77 |

**Effort required at traversing handwheel** (in-lb)

| Left traverse | High gear | Low gear |
|---|---|---|
| 0° | 55 | 39 |
| 90° | 28 | 6 |
| 180° | 11 | 11 |
| 270° | 22 | 17 |

| Right traverse | High gear | Low gear |
|---|---|---|
| 0° | 10 | 6 |
| 90° | 44 | 17 |
| 180° | 61 | 44 |
| 270° | 20 | 17 |

**Time to elevate from -3 to +85°:**
High gear—15.02sec
Low gear—25.90sec
**Time to depress from +85 to -3°:**
High gear—21.44sec
Low gear—34.90sec
**Time to traverse 360°:**
High gear—33.90sec
Low gear—69.79sec
**Over-all dimensions in firing position:**
Length—5.8m (19ft)
Height—2.1m (6.9ft)
Width—5.14m 16.87ft w/outriggers
**Overall dimensions in traveling position:**
Length—7.7m w/drawbar (25.5ft)
Height—2.4m (7.9ft)
Width (front)—2.19m (7.2ft)
Width (rear)—2.3m (7.6ft)
**Length of outriggers:** 1.46m (4.8ft)
**Number of bogies:** 2
**Type of bogies**: Single axle. Single wheels on front; dual wheels on rear
**Weight of front bogie**: 827.8kg (1,825lb)
**Weight of rear bogie**: 1,199.7kg (2,645lb)
**Pneumatic tire size:** 81.2cm x 16.5cm (32in x 6½in) (6½ extra 20) also marked 7:50 x 20
**Wheel base**: 4.19m (13.75ft)
**Type of brakes:** Vacuum air brakes on all wheels; hand-operated parking brakes on rear wheels also
**Type and number of jacks:** 4 jacks integral with mount for leveling bottom carriage; one on each end of outriggers and carriage
**Leveling**: 4.5° leveling either side
**Road clearance:** 34.7cm (1.14ft)
**Tread:** front—176.78cm (5.8ft)
rear—182.8cm (6ft)
**Height of axis of bore above ground (firing):** 152.4cm (5ft)
**Time to change from traveling to firing position:** approx. 2½min with 6-man crew
**Time to change from firing to traveling position:** approx. 3½min with 6-man crew
**Weight of entire carriage:** 7,404.85kg (16,325lb)
**Rear wheel reactions:** 4,458.78kg (9,830lb)

**Front wheel reactions:** 2,952.87kg (6,510lb)
**Type of equilibrators:** spring type with built-in spring compressors

## 4 Ammunition

Rounds for the 8.8cm were of 'fixed' type—the round was supplied with the cartridge joined with the shell. The cartridge length could vary with the type of gun for which it was supplied, for example the Flak 41 had a longer cartridge than earlier Flak models. In general rounds were either HE (high explosive)—in German *Sprenggranate*, shortened to SprGr—or AP (armour piercing)—in German *Panzergranate*, PzGr. The cartridge (*Patrone*) was designated separately from the round.

Originally the round had copper or brass driving bands and a brass cartridge case. War economies led to sintered iron driving bands and steel cartridge cases. Rounds could be nose-fused (*Kopfzundung*, Kz) in the case of HE, or base-fused (*Bodenzunder*, Bdz), usually for AP rounds. Time fuses were normally used for HE, but a percussion fuse was developed by the last year of the war. Numerous different 8.8cm rounds were used but the most common are listed below.

### Ammunition supply

Complete rounds were supplied individually sealed in cylindrical sealed steel containers with clip-sealed top, or in threes in wicker and wood containers with metal base and metal top, the latter held in place by a leather strap. Rubber pads in the base protected the fuses against jarring.

| Round | Weapon | Weight | Colour of shell |
|---|---|---|---|
| 8.8cm SprGr. L/45 | Flak 18,36,37 | 9.24kg | Yellow or green |
| 8.8cm PzGr. 39 | Flak 18-37,39(r) KwK 36 | 10kg | Black or black/red |
| 8.8cmPzGr. 40 | Flak 36, 41, KwK 36, KwK 43, Pak 43 | 7.27kg | Black |
| 8.8cm SprGr. L/4.7 | Flak 41 | 9.4kg | Yellow |
| 8.8cm SprGr. 43 | KwK 43, Pak 43,43/41 | 9.4kg | Olive |
| 8.8cm Lgr. L/4.4 | Flak 18,36,37,41,39(r) | 9.3kg | Green, black tip |

Lgr.=Leichtgranate. This was a starshell round incorporating a parachute flare.

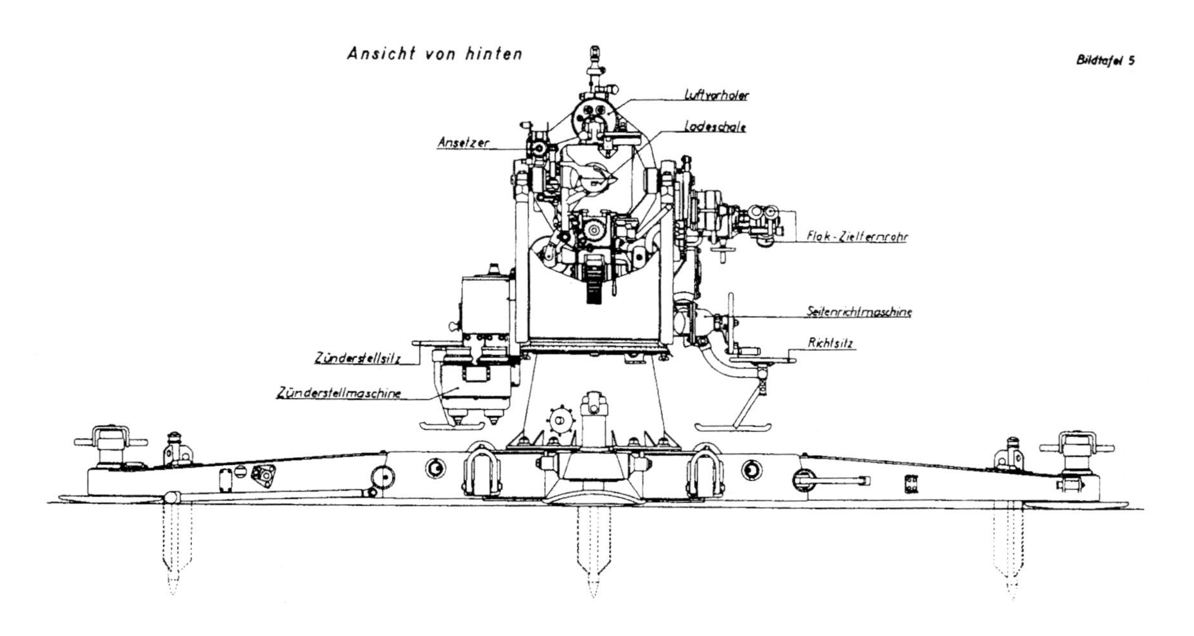


LEFT: German ordnance drawing of the rear view of the Flak 18.

RIGHT: German ordnance drawings of the Flak 18—from top to bottom, left-side view; section down centreline; right-side view; view from above.

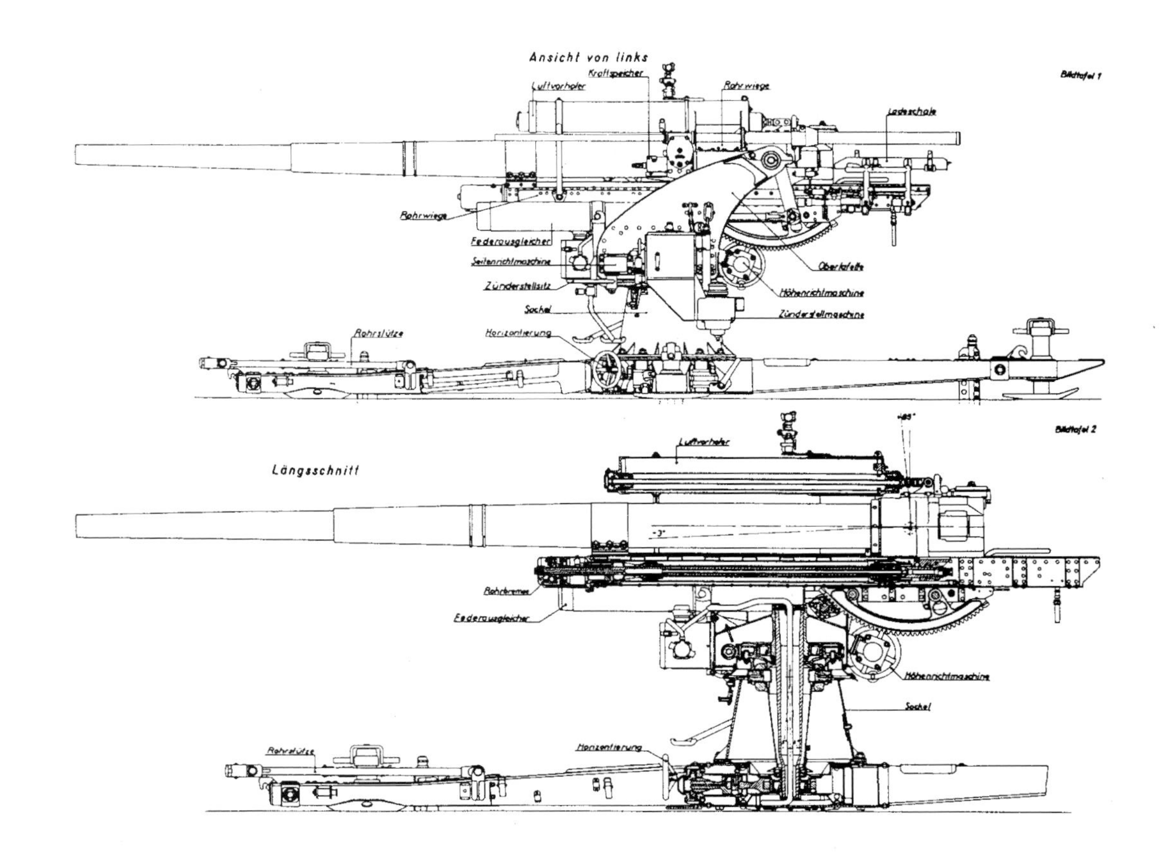


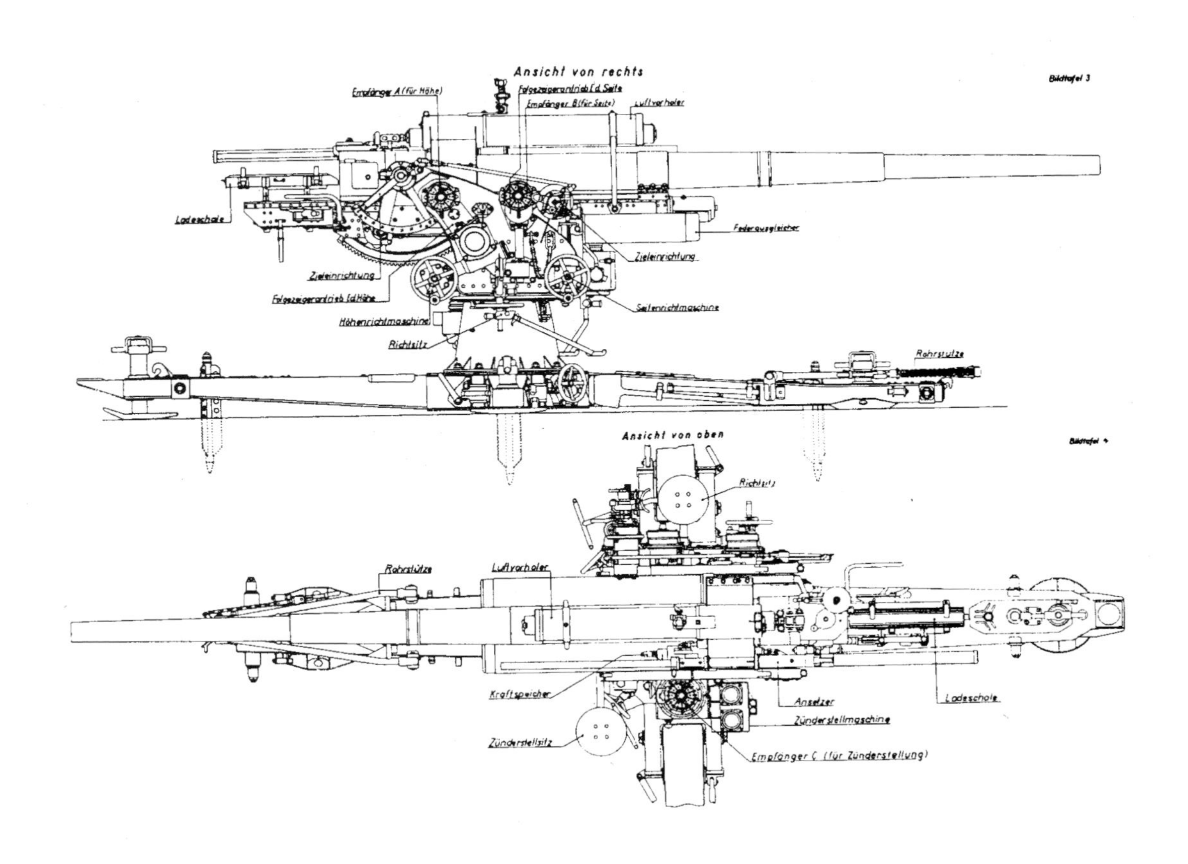